OFFICIAL SQA PAST PAPERS WITH ANSWERS

HIGHER

HUMAN BIOLOGY
2006-2010

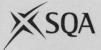

© Scottish Qualifications Authority
All rights reserved. Copying prohibited. No part of this publication may be reproduced, stored in a retrieval system, or transmitted in any form or by any means, electronic, mechanical, photocopying, recording or otherwise.

First exam published in 2006.
Published by Bright Red Publishing Ltd, 6 Stafford Street, Edinburgh EH3 7AU
tel: 0131 220 5804 fax: 0131 220 6710 info@brightredpublishing.co.uk www.brightredpublishing.co.uk

ISBN 978-1-84948-142-7

A CIP Catalogue record for this book is available from the British Library.

Bright Red Publishing is grateful to the copyright holders, as credited on the final page of the book, for permission to use their material.
Every effort has been made to trace the copyright holders and to obtain their permission for the use of copyright material.
Bright Red Publishing will be happy to receive information allowing us to rectify any error or omission in future editions.

[BLANK PAGE]

FOR OFFICIAL USE

Total for
Sections B & C

X009/301

NATIONAL
QUALIFICATIONS
2006

TUESDAY, 23 MAY
1.00 PM – 3.30 PM

HUMAN BIOLOGY
HIGHER

Fill in these boxes and read what is printed below.

Full name of centre

Town

Forename(s)

Surname

Date of birth

Day   Month   Year

Scottish candidate number

Number of seat

SECTION A—Questions 1–30

Instructions for completion of Section A are given on page two.

For this section of the examination you must use an **HB pencil**.

SECTIONS B AND C

1   (a)   All questions should be attempted.

     (b)   It should be noted that in **Section C** questions 1 and 2 each contain a choice.

2   The questions may be answered in any order but all answers are to be written in the spaces provided in this answer book, **and must be written clearly and legibly in ink**.

3   Additional space for answers will be found at the end of the book. If further space is required, supplementary sheets may be obtained from the invigilator and should be inserted inside the **front** cover of this book.

4   The numbers of questions must be clearly inserted with any answers written in the additional space.

5   Rough work, if any should be necessary, should be written in this book and then scored through when the fair copy has been written. If further space is required a supplementary sheet for rough work may be obtained from the invigilator.

6   Before leaving the examination room you must give this book to the invigilator. If you do not, you may lose all the marks for this paper.

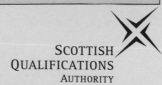

SCOTTISH
QUALIFICATIONS
AUTHORITY

Read carefully

1 Check that the answer sheet provided is for **Human Biology Higher (Section A)**.

2 For this section of the examination you must use an **HB pencil**, and where necessary, an eraser.

3 Check that the answer sheet you have been given has **your name**, **date of birth**, **SCN** (Scottish Candidate Number) and **Centre Name** printed on it.

Do not change any of these details.

4 If any of this information is wrong, tell the Invigilator immediately.

5 If this information is correct, **print** your name and seat number in the boxes provided.

6 The answer to each question is **either** A, B, C or D. Decide what your answer is, then, using your pencil, put a horizontal line in the space provided (see sample question below).

7 There is **only one correct** answer to each question.

8 Any rough working should be done on the question paper or the rough working sheet, **not** on your answer sheet.

9 At the end of the exam, put the **answer sheet for Section A inside the front cover of this answer book**.

Sample Question

The digestive enzyme pepsin is most active in the

A stomach

B mouth

C duodenum

D pancreas.

The correct answer is **A**—stomach. The answer **A** has been clearly marked in **pencil** with a horizontal line (see below).

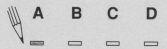

Changing an answer

If you decide to change your answer, carefully erase your first answer and, using your pencil, fill in the answer you want. The answer below has been changed to **D**.

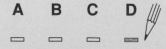

SECTION A

All questions in this section should be attempted.
Answers should be given on the separate answer sheet provided.

1. The diagram below shows a mitochondrion surrounded by cytoplasm.

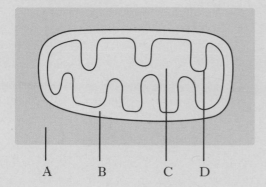

   Where does glycolysis take place?

2. Which of the following statements refer to glycolysis?

   1   Carbon dioxide is released.

   2   Occurs during aerobic respiration.

   3   The end product is pyruvic acid.

   4   The end product is lactic acid.

   A   1 and 3

   B   1 and 4

   C   2 and 3

   D   2 and 4

3. In respiration, the products of the cytochrome system are

   A   hydrogen and carbon dioxide

   B   water and ATP

   C   oxygen and ADP

   D   pyruvic acid and water.

4. During anaerobic respiration in muscle fibres, what is the fate of pyruvic acid?

   A   It is converted to lactic acid.

   B   It is broken down by the mitochondria.

   C   It is broken down to carbon dioxide and water.

   D   It is converted to citric acid.

5. The table below shows the antigens and antibodies present in the four different blood groups of the ABO system.

Group	Antigen	Antibody
1	B	a
2	none	a and b
3	A and B	none
4	A	b

   Which of these groups could safely receive a transfusion of blood of group A?

   A   1 and 2

   B   1 and 4

   C   2 and 3

   D   3 and 4

6. Which of the following is a cell that engulfs bacteria?

   A   B-lymphocyte

   B   T-lymphocyte

   C   Lysosome

   D   Macrophage

7. Which of the following processes occurs during the second division of meiosis?

   A   Formation of diploid daughter cells

   B   Pairing of homologous chromosomes

   C   Separation of paired chromatids

   D   Crossing over of genetic material

8. Polygenic characteristics are different from monohybrid characteristics because they

   A   show random assortment of chromosomes

   B   show independent assortment of chromosomes

   C   are controlled by many pairs of alleles

   D   are caused by non-disjunction during meiosis.

9. The gene (m) which causes one type of muscular dystrophy is sex-linked and recessive to the normal gene (M). If a carrier female and an unaffected male have children, what would be the predicted effect on their sons and daughters?

	Sons	Daughters
A	100% are affected	100% are carriers
B	50% are affected	50% are carriers
C	50% are affected	100% are carriers
D	100% are affected	50% are carriers

10. Red-green colour deficient vision is a sex-linked condition. John, who is affected, has the family tree shown below.

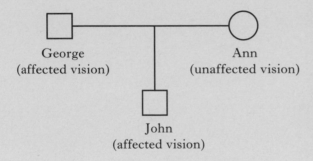

George
(affected vision)

Ann
(unaffected vision)

John
(affected vision)

If b is the mutant allele for the condition, which of the following could be the genotypes of George's parents and Ann's parents?

	George's parents		Ann's parents	
A	X^BX^b	X^BY	X^BX^B	X^BY
B	X^BX^B	X^bY	X^BX^B	X^BY
C	X^BX^b	X^BY	X^BX^b	X^BY
D	X^BX^B	X^bY	X^BX^B	X^bY

11. After ovulation, the follicle develops into the

A corpus luteum

B fallopian tube

C endometrium

D zygote.

12. Which of the following best describes monozygotic twins?

A They are genetically similar and have developed from two eggs fertilised by two sperm.

B They are genetically similar and have developed from one egg fertilised by two sperm.

C They are genetically identical and have developed from one egg fertilised by one sperm.

D They are genetically identical and have developed from one egg fertilised by two sperm.

13. Which of the following sequences describes the first stages in the development of an embryo?

A fertilisation → cleavage → implantation

B implantation → fertilisation → cleavage

C cleavage → fertilisation → implantation

D fertilisation → implantation → cleavage

14. Which of the following hormones is produced by the placenta?

A Growth hormone

B Prolactin

C Progesterone

D Oxytocin

15. The graph shows changes in lung volume during a breathing exercise.

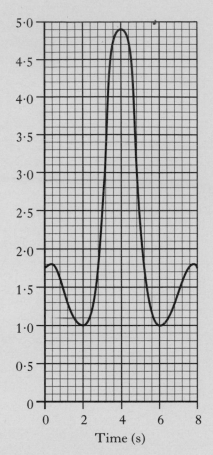

Lung volume (litres)

Time (s)

What is the volume of air exhaled between 4 and 6 seconds?

A   3·8 litres

B   3·9 litres

C   4·8 litres

D   4·9 litres

16. Which of the following structures is **not** involved in the production or breakdown of red blood cells?

A   Spleen

B   Pancreas

C   Liver

D   Bone marrow

17. The diagram shows a cross-section of the heart.

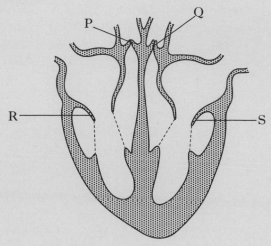

Which of the following describes correctly the movement of the valves during ventricular systole?

A   Valves P and Q open and valves R and S close

B   Valves P and R open and valves Q and S close

C   Valves P and Q close and valves R and S open

D   Valves P and R close and valves Q and S open

18. The trace below was obtained from a patient who was having the electrical activity of his heart monitored.

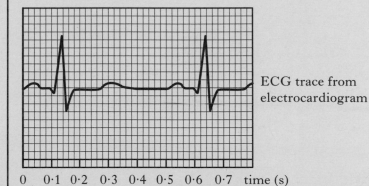

ECG trace from electrocardiogram

time (s)

What was the heart rate of this patient?

A   42 beats per minute

B   72 beats per minute

C   86 beats per minute

D   120 beats per minute

19. The graph below shows how pulse rate and stroke volume change with the rate of oxygen uptake.

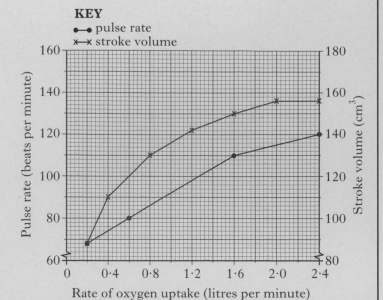

KEY
●—● pulse rate
×—× stroke volume

Cardiac output = pulse rate × stroke volume

What is the cardiac output when the oxygen uptake is 1·6 litres per minute?

A   13·1 litres per minute

B   14·3 litres per minute

C   16·5 litres per minute

D   16·9 litres per minute

20. In a healthy human, blood entering the kidney contains more glucose than blood leaving the kidney because the glucose is

A   changed to waste by the kidney tubules

B   stored by the kidney cells

C   excreted by the kidney tubules

D   used by the kidney cells for respiration.

21. The concentration of urea rises from $0.03 \, g/100 \, cm^3$ to $0.15 \, g/100 \, cm^3$ as it passes through a kidney tubule.

What is the difference in concentration, expressed as a whole number ratio?

A   1 : 5

B   1 : 50

C   3 : 100

D   2 : 1

22. When body temperature rises, which of the following is true of blood flow in the skin capillaries?

A   The flow of blood in the capillaries increases and heat loss decreases.

B   The flow of blood in the capillaries increases and heat loss increases.

C   The flow of blood in the capillaries decreases and heat loss decreases.

D   The flow of blood in the capillaries decreases and heat loss increases.

23. In which part of the brain are the control centres for both speech and hearing located?

A   Limbic system

B   Hypothalamus

C   Medulla oblongata

D   Cerebrum

24. The function of the corpus callosum is to

A   transfer information from a sensory nerve to a motor nerve

B   control balance and coordination

C   transfer information from one hemisphere to the other

D   control all sensory activities.

25. In which of the following is part of the autonomic nervous system correctly linked to the response it causes?

	Part of the autonomic nervous system	Response
A	sympathetic	acceleration of heart beat
B	sympathetic	vasodilation of skin arterioles
C	parasympathetic	secretion of sweat
D	parasympathetic	vasodilation of coronary blood vessels

26. When a person's beliefs are changed as a result of persuasion, this is an example of

A   internalisation

B   identification

C   deindividuation

D   social facilitation.

27. The graph below contains information about fertiliser usage.

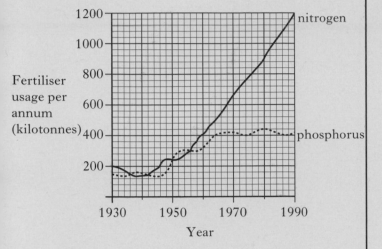

Which of the following statements about nitrogen usage between 1930 and 1990 is correct?

A   It increased steadily.

B   It increased by 500%.

C   It increased by 600%.

D   It always exceeded phosphorus usage.

28. The bar chart below shows the percentage loss in yield of four organically grown crops as a result of the effects of weeds, disease and insects.

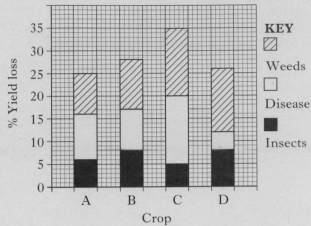

Predict which crop is most likely to show the greatest increase in yield if herbicides and insecticides were applied.

[Turn over

29. The graph below contains information about the birth rate and death rate in Mexico.

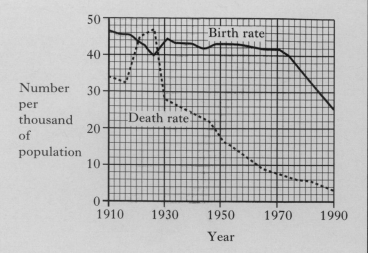

Which of the following conclusions can be drawn from the graph?

A   At no time during the century has the population of Mexico decreased.

B   The greatest increase in population occurred in 1970.

C   The population was growing faster in 1910 than in 1990.

D   Birth rate decreased between 1970 and 1990 due to the use of contraception.

30. The diagram below shows the carbon cycle.

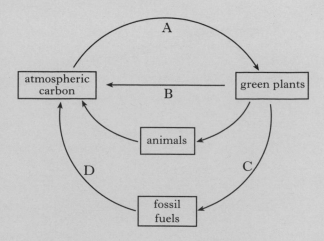

Which letter represents combustion?

Candidates are reminded that the answer sheet MUST be returned INSIDE the front cover of this answer booklet.

DO NOT
WRITE IN
THIS
MARGIN

Marks

SECTION B

All questions in this section should be attempted.

All answers must be written clearly and legibly in ink.

1. (*a*) The diagram below shows a structural model of the plasma membrane.

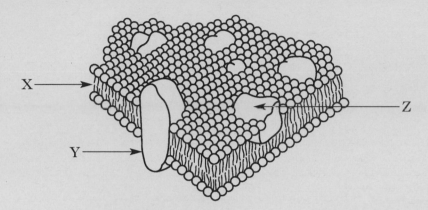

   (i) What term describes this model of the membrane?

_____   1

   (ii) Identify components X and Y.

X _____

Y _____   1

   (iii) State a possible function of Z.

_____   1

(*b*) Sodium ions can be moved against a concentration gradient across a membrane.

   (i) Explain what is meant by a concentration gradient across a membrane.

_____

_____   1

   (ii) What term describes the movement of ions against a concentration gradient?

_____   1

   (iii) Explain why a shortage of oxygen might lead to a decrease in the rate of sodium ion movement.

_____

_____

_____   2

Marks

2. (*a*) Complete the table below to show the mRNA codons and tRNA anticodons for each amino acid.

Amino acid	mRNA codons	tRNA anticodons
alanine		CGA
threonine	ACC	
cysteine		ACA

1

(*b*) The diagram shows the primary structure of part of a protein molecule.

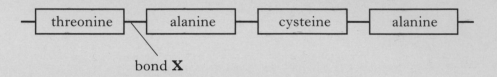

bond **X**

(i) Use the information from the table above to determine the DNA base sequence which would code for this molecule.

_____   **1**

(ii) Name bond **X**.

_____   **1**

(iii) Describe **one** way in which the secondary structure of a protein differs from the primary structure.

_____

_____

_____

_____   **1**

(*c*) Where in the cell are proteins packaged and prepared immediately before secretion?

_____   **1**

Marks

3.   The diagram shows a polio virus.

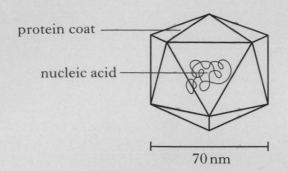

protein coat

nucleic acid

70 nm

(a)   Viruses can only reproduce within a host cell.

     (i)   List **two** substances, supplied by the host cell, which are required for viral replication.

        1 _____

        2 _____   1

    (ii)   What happens after the viruses have been assembled inside the cell?

        _____

        _____

        _____   1

(b)   Viruses can be processed to make vaccines to protect against the disease.

     Suggest why it is important that the nucleic acid is damaged in the process, but not the protein coat.

     Nucleic acid damaged _____

    _____

    _____   1

     Protein coat undamaged _____

    _____

    _____   1

(c)   The average diameter of a red blood cell is 7 µm.

     By how many times is a red blood cell bigger than a polio virus? (1 µm = 1000 nm)

     Space for calculation

                                           1

DO NO
WRITE
THIS
MARG

4. Polydactyly is an inherited condition in which individuals are born with extra toes. *Marks*
The allele for polydactyly is dominant and not sex-linked.

The family tree below shows the incidence of the condition through three generations.

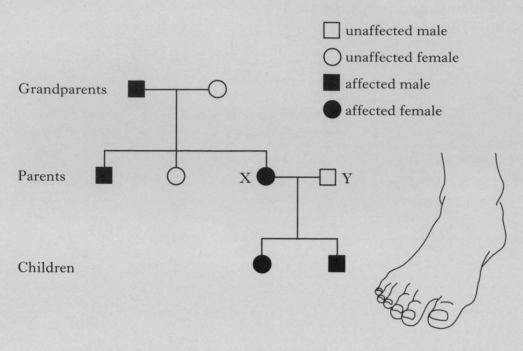

□ unaffected male

○ unaffected female

■ affected male

● affected female

Grandparents

Parents

Children

(a)   (i)   Using the symbol **D** for the allele for polydactyly and **d** for the normal allele, give the genotypes of the two children.

_____ and _____   1

(ii)   Individuals X and Y are expecting another child.

What are the chances of this child inheriting the condition?

_____   1

(b)   (i)   What evidence from the family tree confirms that the grandfather is heterozygous?

_____

_____   1

(ii)   What evidence from the family tree confirms that the condition is not sex-linked?

_____

_____   1

(c)   What term is used to refer to chromosomes which are not sex-chromosomes?

_____   1

DO NOT WRITE IN THIS MARGIN

Marks

5. The diagrams show the hormonal control of the testes and ovaries by the pituitary gland.

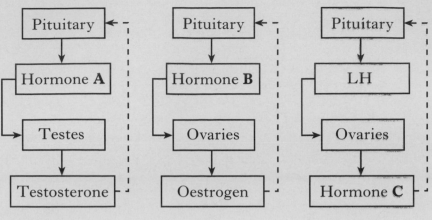

$\longrightarrow$ stimulation $- - - - - \rightarrow$ inhibition

(*a*) (i) What name is given to this type of hormonal control?

_____ 1

(ii) Identify hormones **A**, **B** and **C**.

A _____

B _____

C _____ 2

(iii) State an effect of oestrogen on the pituitary gland, other than that shown above.

_____ 1

(iv) Where in the testes is testosterone produced?

_____ 1

(*b*) Distinguish between cyclical fertility and continuous fertility.

_____

_____

_____ 1

(*c*) The female contraceptive pill raises the levels of ovarian hormones in the blood. Explain why this has a contraceptive effect.

_____

_____

_____ 2

DO NOT
WRITE
THIS
MARGIN

6. The diagram shows the blood supply between a fetus and its placenta.

Marks

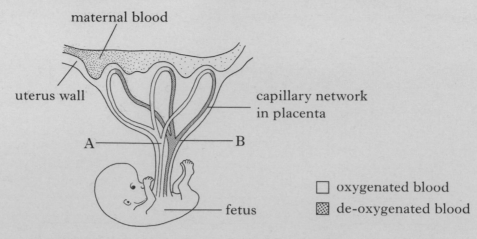

maternal blood

uterus wall

capillary network
in placenta

A   B

fetus

☐ oxygenated blood
▨ de-oxygenated blood

(a) Name **two** waste products that pass from the fetal blood to the maternal blood.

1 _____ 2 _____   1

(b) The table shows some substances and their method of exchange between the fetal and maternal blood. Complete the table.

Substance	Method of exchange
	diffusion
glucose	
antibodies	

2

(c) Which of the fetal blood vessels, A or B, is the artery?
Give a reason for your answer.

Vessel _____

Reason for answer _____

_____   1

(d) Why might the second Rhesus positive child of a Rhesus negative mother be in danger from the mother's immune system?

_____

_____

_____

_____   2

(e) Why do some inborn errors of metabolism, such as phenylketonuria (PKU), only have an effect on the baby *after* birth?

_____

_____

_____   1

Marks

7. The diagram shows the liver and its associated blood supply.

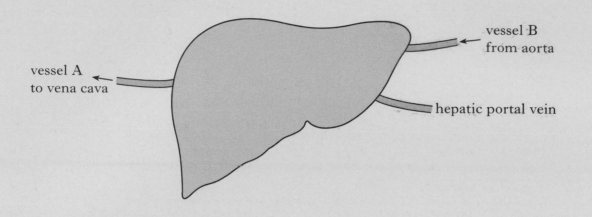

vessel B
from aorta

vessel A
to vena cava

hepatic portal vein

(*a*) (i) Identify the vessels labelled A and B.

A _____   B _____   1

(ii) Name an organ which is linked to the liver by the hepatic portal vein.

_____   1

(*b*) The table below relates to products of digestion, their vessel of transportation
in the villus and their possible fate in the body.

Complete the table.

Product of digestion	Vessel of transportation	Possible fate
glucose		
		deamination in the liver
	lacteal	

3

(*c*) Identify **two** hormones which cause the liver to release glucose, and state the
conditions under which each of the hormones is released.

1 Hormone _____

Condition _____   1

2 Hormone _____

Condition _____   1

[Turn over

DO NOT
WRITE I
THIS
MARGI

Marks

8. A student carried out an investigation into the effectiveness of thermal insulation.

Two flasks containing water at 50 °C were left for forty minutes. During this time the temperature of the water was recorded every ten minutes. The results are given in the table below.

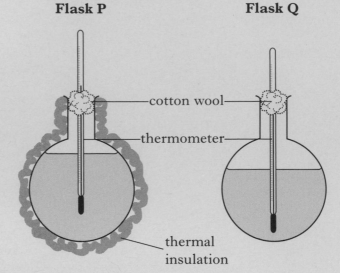

Flask P   **Flask Q**

	Temperature (°C)	
Time (min)	Flask P	Flask Q
0	50	50
10	44	37
20	39	30
30	36	24
40	34	20

cotton wool

thermometer

thermal insulation

(a) Present the data in a suitable form on the graph paper.

(Additional graph paper, if required, can be found on page 28.)

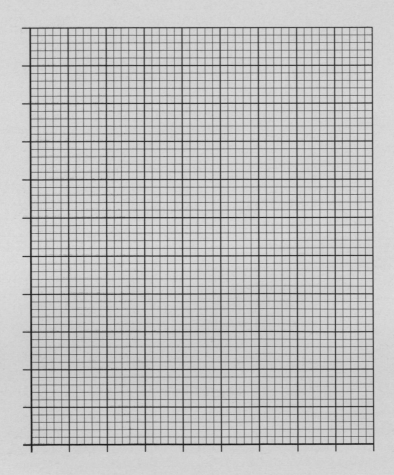

3

Marks

8. **(continued)**

   (b) Calculate the percentage decrease in temperature in flasks P and Q over the 40-minute period.

   P_____ % Q _____ %

   1

   (c) The student concluded that the insulation had slowed the cooling rate of the flask.

   Describe **two** aspects of the experimental design which make his conclusion invalid.

   1 _____

   _____

   2 _____

   _____

   1

   (d) Another student went on to compare flasks of different sizes without any insulation. She compared a $50\,cm^3$ flask with a $100\,cm^3$ flask, each completely filled with hot water.

   (i) State **two** variables that would have to be kept the same during this second investigation.

   1 _____

   2 _____

   1

   (ii) Which flask would cool more quickly? Give a reason for your answer.

   Flask _____

   Reason _____

   _____

   _____

   1

   (e) What part of the brain monitors body temperature?

   _____

   1

   [Turn over

Marks

9. The diagram shows a neuromuscular synapse.

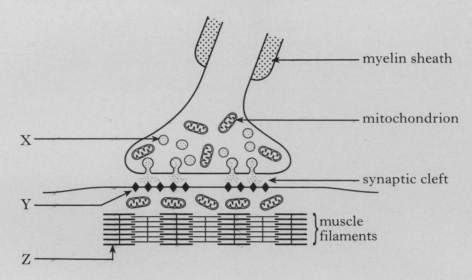

myelin sheath

mitochondrion

X

synaptic cleft

Y

{muscle filaments

Z

(*a*)　(i)　Name cell structure **X**.

_____　1

(ii)　Describe the role of structure **X** in exocytosis.

_____

_____　1

(*b*)　What is the function of molecule **Y**?

_____

_____　1

(*c*)　The areas on both sides of the synaptic cleft are rich in mitochondria. Explain why mitochondria are needed in each area.

_____

_____

_____　2

(*d*)　(i)　Name protein filament **Z**.

_____　1

(ii)　Describe what happens to the length of this filament when the muscle contracts.

_____　1

Marks

10. An investigation was carried out to study the serial position effect.

Twelve pictures were shown, one by one, to five children.

The children were then asked to recall the pictures they saw.

The results of the investigation are shown below.

The table shows the recall success for each picture.

Child	Position of picture in list shown to children											
	1st	2nd	3rd	4th	5th	6th	7th	8th	9th	10th	11th	12th
1	✓	✓	✓	✓	✗	✗	✓	✗	✓	✓	✓	✓
2	✓	✓	✓	✗	✗	✓	✗	✗	✓	✗	✓	✓
3	✓	✗	✓	✓	✗	✗	✗	✗	✗	✓	✓	✗
4	✓	✓	✗	✗	✗	✗	✓	✓	✓	✓	✓	✓
5	✓	✓	✓	✗	✓	✗	✗	✓	✗	✓	✓	✓
Recall (%)	100	80	80	40	20	20	40	40	60	80	100	80

✓ = picture recalled   ✗ = picture forgotten

(*a*)   (i)   Describe the trend shown by these results.

_____

_____

_____   **1**

(ii)   Explain these results in terms of the serial position effect.

_____

_____

_____

_____

_____

_____

_____

_____

_____   **3**

(*b*)   To make sure that the children tried their best, the investigation was designed as a competition and the child with the best recall was rewarded.

What behavioural term describes improved performance in competitive situations?

_____   **1**

DO NO
WRITE
THIS
MARG

Marks

11. Analysis of fertility rates can be used to predict population change over the next hundred years.

Graph 1 shows the number of children born to a sample of twenty UK women of reproductive age. The sample was taken in the year 2000.

Graph 1

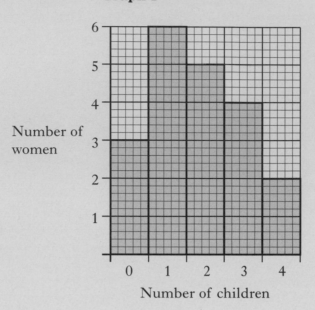

Number of women

Number of children

(a) The fertility rate for a country is calculated by dividing the total number of children by the number of women in the sample.

(i) From **Graph 1**, calculate the fertility rate of this UK sample.

_____

1

(ii) How could the calculation of the UK fertility rate be made more reliable?

_____

_____

_____

1

(iii) The age of each woman is not given. Why might this information be important?

_____

_____

_____

1

Marks

11.   **(continued)**

Graph 2 shows the predicted population changes in the UK for four different fertility rates.

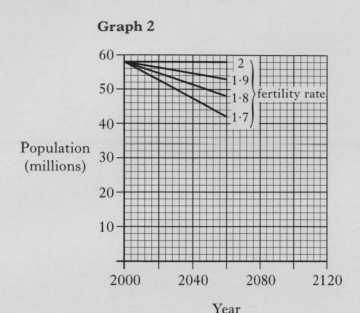

Graph 2

(b)   Using the fertility rate you calculated from **Graph 1**, predict the population of the UK in the year 2100.

_____   1

(c)   In a sample of twenty families in Thailand, a developing country, three women have two children, ten women have three children and seven women have four children.

   (i)   What is the fertility rate for this sample?
          Space for calculation

                                                                    _____   1

   (ii)   Suggest a reason for the difference in fertility rate between Thailand and the UK.

   _____

   _____   1

(d)   Birth rates and death rates can also be used to predict population changes. Give **one** other factor which would affect the size of a population.

_____   1

(e)   What term is used to describe studies of population statistics such as this?

_____   1

DO NO
WRITE
THIS
MARGI

12. The tables below contain information about the population of the United Kingdom in the year 2000.

Marks

Table 1 – Populations of individual countries

Country	Population (millions)
England	48·9
Scotland	4·9
Wales	2·7
Nothern Ireland	1·5
Total	**58·0**

Table 2 – Population profile of UK

Group	Numbers (millions)
Under 16 years	11·6
16–59 years	34·4
60 years and over	12·0
Males	28·0
Females	30·0

(*a*) From **Table 1**, calculate the percentage of the UK population that is Scottish.

Space for calculation

———————— %   **1**

(*b*) From **Table 2**, calculate the male to female sex ratio.

Space for calculation

————— : —————   **1**
male   female

(*c*) Use the information in **Tables 1** and **2** to estimate the number of children under sixteen years of age, living in Scotland.

Space for calculation

————————   **1**

Marks

13. The diagram below shows three possible fates of nitrates which have been added to the soil as fertiliser.

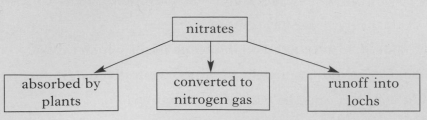

(*a*) Why are nitrates essential for plant growth?

_____   1

(*b*) What type of bacteria convert nitrate to nitrogen gas?

_____   1

(*c*) Explain how the runoff of nitrates into a loch ecosystem might result in a drop in the oxygen concentration of the water.

_____

_____

_____

_____

_____   2

(*d*) Describe **two** ways by which the nitrate content of the soil can increase naturally.

1 _____

_____

2 _____

_____   1

[Turn over for SECTION C on *Page twenty-four*

DO NOT
WRITE
THIS
MARGIN

Marks

SECTION C

Both questions in this section should be attempted.

Note that each question contains a choice.

Questions 1 and 2 should be attempted on the blank pages which follow.

Supplementary sheets, if required, may be obtained from the invigilator.

Labelled diagrams may be used where appropriate.

1. Answer **either** A **or** B.

    A. Give an account of respiration under the following headings:

        (i) the role of ATP within the cell;   **4**

        (ii) the use of different respiratory substrates.   **6**

    (10)

    OR

    B. Give an account of enzymes under the following headings:

        (i) factors affecting enzyme activity;   **7**

        (ii) activation of enzymes.   **3**

    (10)

In question 2, ONE mark is available for coherence and ONE mark is available for relevance.

2. Answer **either** A **or** B.

    A. Describe the effect of experience on learning.   **(10)**

    OR

    B. Discuss the impact of an increasing population on the world's water supplies.   **(10)**

[END OF QUESTION PAPER]

DO NOT WRITE IN THIS MARGIN

SPACE FOR ANSWERS

SPACE FOR ANSWERS

DO NOT
WRITE IN
THIS
MARGIN

SPACE FOR ANSWERS

DO NOT
WRITE IN
THIS
MARGIN

SPACE FOR ANSWERS

ADDITIONAL GRAPH PAPER FOR QUESTION 8(*a*)

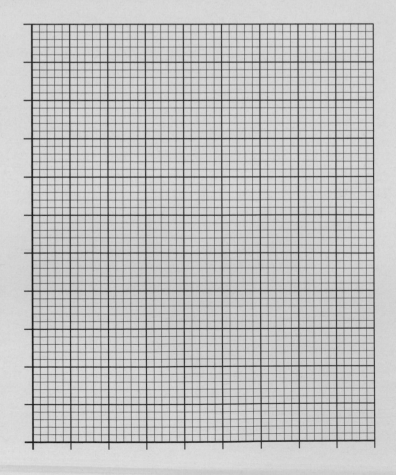

HIGHER

2007

[BLANK PAGE]

FOR OFFICIAL USE

Total for
Sections B & C

X009/301

NATIONAL
QUALIFICATIONS
2007

MONDAY, 21 MAY
1.00 PM – 3.30 PM

HUMAN BIOLOGY
HIGHER

Fill in these boxes and read what is printed below.

Full name of centre

Town

Forename(s)

Surname

Date of birth

Day Month Year

Scottish candidate number

Number of seat

SECTION A—Questions 1–30

Instructions for completion of Section A are given on page two.

For this section of the examination you must use an **HB pencil**.

SECTIONS B AND C

1   (a) All questions should be attempted.

(b) It should be noted that in **Section C** questions 1 and 2 each contain a choice.

2   The questions may be answered in any order but all answers are to be written in the spaces provided in this answer book, **and must be written clearly and legibly in ink**.

3   Additional space for answers will be found at the end of the book. If further space is required, supplementary sheets may be obtained from the invigilator and should be inserted inside the **front** cover of this book.

4   The numbers of questions must be clearly inserted with any answers written in the additional space.

5   Rough work, if any should be necessary, should be written in this book and then scored through when the fair copy has been written. If further space is required a supplementary sheet for rough work may be obtained from the invigilator.

6   Before leaving the examination room you must give this book to the invigilator. If you do not, you may lose all the marks for this paper.

SCOTTISH
QUALIFICATIONS
AUTHORITY

©

Read carefully

1   Check that the answer sheet provided is for **Human Biology Higher (Section A)**.

2   For this section of the examination you must use an **HB pencil** and, where necessary, an eraser.

3   Check that the answer sheet you have been given has **your name**, **date of birth**, **SCN** (Scottish Candidate Number) and **Centre Name** printed on it.

Do not change any of these details.

4   If any of this information is wrong, tell the Invigilator immediately.

5   If this information is correct, **print** your name and seat number in the boxes provided.

6   The answer to each question is **either** A, B, C or D. Decide what your answer is, then, using your pencil, put a horizontal line in the space provided (see sample question below).

7   There is **only one correct** answer to each question.

8   Any rough working should be done on the question paper or the rough working sheet, **not** on your answer sheet.

9   At the end of the exam, put the **answer sheet for Section A inside the front cover of this answer book**.

Sample Question

The digestive enzyme pepsin is most active in the

A   stomach

B   mouth

C   duodenum

D   pancreas.

The correct answer is **A**—stomach. The answer **A** has been clearly marked in **pencil** with a horizontal line (see below).

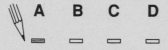

Changing an answer

If you decide to change your answer, carefully erase your first answer and, using your pencil, fill in the answer you want. The answer below has been changed to **D**.

A   B   C   D

SECTION A

All questions in this section should be attempted.

Answers should be given on the separate answer sheet provided.

1. Which line in the table correctly identifies the two cell structures shown in the diagram?

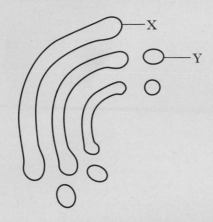

	X	Y
A	Endoplasmic reticulum	Vesicle
B	Endoplasmic reticulum	Ribosome
C	Golgi body	Vesicle
D	Golgi body	Ribosome

2. Which of the following correctly describes metabolism?

   A   The breakdown of chemicals to release energy

   B   The rate at which an organism produces heat energy

   C   The chemical reactions of organisms

   D   The breakdown of food molecules

3. Phenylketonuria (PKU) is a metabolic disorder which can be lethal in childhood. It is caused by an inability to make *enzyme X*, shown in the metabolic pathway below.

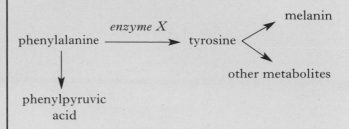

   Which substance would have to be removed from the diet for someone who has this disorder?

   A   Phenylalanine

   B   Enzyme X

   C   Tyrosine

   D   Melanin

4. A stock solution has a concentration of 1 M. $100 \, cm^3$ of a $0.4 \, M$ solution can be prepared by adding

   A $40 \, cm^3$ of stock solution to $60 \, cm^3$ of water

   B $60 \, cm^3$ of stock solution to $40 \, cm^3$ of water

   C $40 \, cm^3$ of stock solution to $100 \, cm^3$ of water

   D $100 \, cm^3$ of stock solution to $40 \, cm^3$ of water.

5. Non-competitive inhibitors affect enzyme action by

   A   acting as a co-enzyme for enzyme action

   B   altering the shape of the substrate molecule

   C   competing for the active site of the enzyme

   D   altering the shape of the active site of the enzyme.

[Turn over

6. The graph shows the effect of substrate concentration on the rate of an enzyme-catalysed reaction.

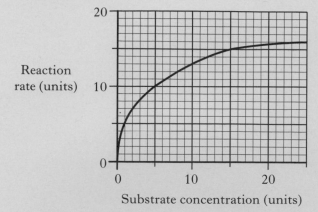

At what substrate concentration is the reaction rate equal to 75% of the maximum rate?

A   6 units

B   8 units

C   12 units

D   18 units

7. Which of the following is **not** a protein?

A   Actin

B   Insulin

C   Amylase

D   Ribonucleic acid

8. The phospholipid molecules in a cell membrane allow the

A   free passage of glucose molecules

B   self-recognition of cells

C   active transport of ions

D   membrane to be fluid.

9. Red blood cells have a solute concentration of around 0·9%.

Which of the following statements correctly describes the fate of these cells when immersed in a 1% salt solution?

A   The cells will burst.

B   The cells will shrink.

C   The cells will expand but not burst.

D   The cells will remain unaffected.

10. The secretion of amylase from a cell is an example of

A   endocytosis

B   exocytosis

C   pinocytosis

D   phagocytosis.

11. Lymphocytes act in the defence of the body by

A   ingesting toxins

B   ingesting pathogens

C   producing lysosomes

D   producing antibodies.

12. The graphs below show the effect of two injections of an antigen on the formation of an antibody.

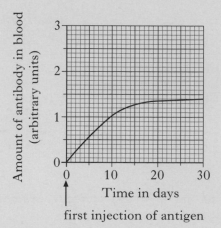

first injection of antigen

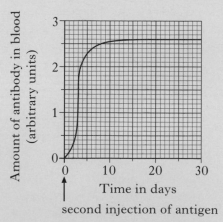

second injection of antigen

How many days after the second injection does the amount of antibody in the blood reach the maximum achieved after the first injection?

A   3 days

B   6 days

C   20 days

D   30 days

13. Haploid gametes are produced during meiosis as a result of

   A   the separation of homologous chromosomes

   B   the independent assortment of chromosomes

   C   the separation of chromosomes into chromatids

   D   the crossing over of chromatids.

14. The diagram refers to human reproduction.

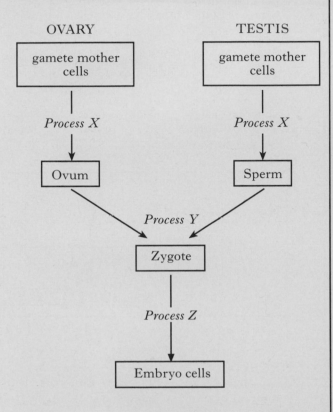

   Which of the following correctly identifies processes X, Y and Z?

	X	Y	Z
A	mitosis	meiosis	fertilisation
B	meiosis	fertilisation	mitosis
C	meiosis	mitosis	fertilisation
D	mitosis	fertilisation	meiosis

15. The family tree shows the pattern of inheritance of a genetic condition.

   Unaffected female   ×   Unaffected male

   ↓

   Affected female

   The allele responsible for this condition is both

   A   sex-linked and recessive

   B   sex-linked and dominant

   C   autosomal and recessive

   D   autosomal and dominant.

16. Non-disjunction can be described as

   A   a metabolic disorder

   B   a type of antisocial behaviour

   C   a condition resulting in memory loss

   D   a form of chromosome mutation.

17. Which of the following organs monitors body temperature?

   A   Hypothalamus

   B   Pituitary gland

   C   Prostate gland

   D   Spleen

18. Which line of the table correctly identifies the function and site of production of bile salts?

	Function	Site of production
A	digest protein	liver
B	digest protein	gall bladder
C	emulsify fats	liver
D	emulsify fats	gall bladder

[Turn over

19. Which of the following vessels in the circulatory system contains blood at the lowest pressure?

    A　Jugular vein

    B　Renal vein

    C　Vena cava

    D　Hepatic portal vein

20. The following data refer to the breathing of an athlete (a) resting and (b) just after a race.

	Breathing rate (breaths per minute)	Volume of one breath	Carbon dioxide in exhaled air (%)
(a) Resting	10	500 ml	5
(b) After race	22	1 litre	5

    Assuming the rate of breathing remains constant, what would be the volume of carbon dioxide breathed out during the first two minutes after the race?

    A　1·1 litres

    B　2·2 litres

    C　22 litres

    D　44 litres

21. The table shows the masses of various substances in the glomerular filtrate and in the urine over a period of 24 hours.

    Which of the substances has the smallest percentage reabsorption from the glomerular filtrate?

	Substance	Mass in glomerular filtrate (g)	Mass in urine (g)
A	Sodium	600·0	6·0
B	Potassium	35·0	2·0
C	Uric acid	8·5	0·8
D	Calcium	5·0	0·2

22. Which of the following shows the direction of a nerve impulse in a neurone?

    A　Axon → cell body → dendrite

    B　Cell body → dendrite → axon

    C　Cell body → axon → dendrite

    D　Dendrite → cell body → axon

23. The diagram below shows the ages (in months) at which children reach various stages in their development.

    The left end of each bar indicates the age by which 25% of infants have reached the stated performance.

    The right end of each bar indicates the age by which 90% of infants have reached the stated performance.

    The vertical bar indicates the age by which 50% of infants have reached the stated performance.

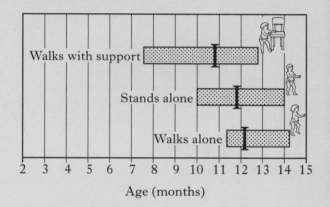

    Age (months)

    An eight-month old infant can walk with support but cannot stand alone.

    In what percentage of the population is this child found?

    A　Less than 25%

    B　Between 25% and 50%

    C　Around 50%

    D　Between 50% and 90%

24. Identical twins are valuable in the study of behaviour because

    A　genetic and environmental factors can be discounted

    B　maturation and environmental factors can be discounted

    C　genetic factors can be discounted

    D　genetic, maturation and environmental factors can be discounted.

25. Which of the following terms describes the process by which a person learns to distinguish between different but related stimuli?

   A   Generalisation

   B   Imitation

   C   Discrimination

   D   Identification

26. An investigation was carried out to determine how long it takes students to learn to run a finger maze. A blindfolded student was allowed to run the maze on ten occasions. The results are given in the table below.

Trial	Time (s)
1	23
2	20
3	26
4	12
5	18
6	10
7	6
8	7
9	6
10	6

How could the investigation be improved to make the results more reliable?

   A   Allow other students to try to run the maze ten times, whilst blindfolded

   B   Allow the same student some additional trials on the same maze

   C   Change the shape of the maze and allow the same student to repeat ten trials

   D   Record the times to one decimal place

27. Which of the following is a correct definition of demography?

   A   Calculation of the difference between birth rates and death rates

   B   A count of the number of individuals in a population

   C   The rate at which a population replaces itself

   D   The study of population numbers

28. Which of the following processes increases directly the concentration of nitrogen gas in the atmosphere?

   A   Decomposition

   B   Denitrification

   C   Detoxification

   D   Deamination

29. The diagram below shows part of the nitrogen cycle.

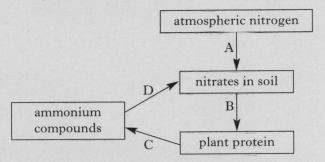

Which letter represents nitrogen fixation?

30. Over-application of which of the following substances on agricultural land is likely to induce algal blooms in adjacent lakes?

   A   Fertiliser

   B   Insecticide

   C   Fungicide

   D   Herbicide

Candidates are reminded that the answer sheet MUST be returned INSIDE the front cover of this answer booklet.

[Turn over for Section B

[BLANK PAGE]

Marks

SECTION B

All questions in this section should be attempted.

All answers must be written clearly and legibly in ink.

1. The diagram below shows part of a DNA molecule.

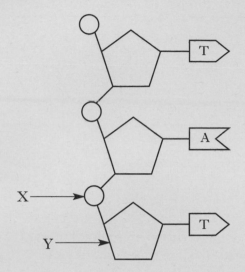

(*a*) (i) On the diagram, draw a circle around **one** nucleotide.   1

    (ii) Name parts X and Y.

       X _____

       Y _____   1

(*b*) Name the **two** DNA bases **not** shown in the diagram.

_____ and _____   1

(*c*) (i) State the mRNA codon which would be formed from the triplet of DNA bases shown.

       _____   1

    (ii) Apart from nucleotides, name another molecule needed for the synthesis of mRNA.

       _____   1

(*d*) A DNA molecule was found to contain 15 000 nucleotides.

What is the maximum number of amino acids which could be coded for by this molecule?

_____   1

[Turn over

Marks

2.   The diagram below shows three stages that occur during aerobic respiration.

| **Stage X** Reactions in the cytoplasm | 1 → | **Stage Y** Reactions in matrix of mitochondrion | 2 → | **Stage Z** Reactions in cristae of mitochondrion | → metabolic products |

(*a*)   Name each stage.

X _____

Y _____

Z _____   **2**

(*b*)   (i)   Arrows 1 and 2 represent the transfer of molecules from one stage to another. Complete the table to identify these molecules.

Arrow	*Name of molecule*
1	
2	

2

(ii)   Name the **two** metabolic products of stage Z.

_____ and _____   **1**

(*c*)   The diagram below shows a mitochondrion from a skin cell.

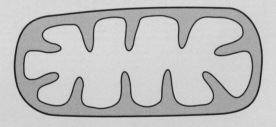

Describe how the structure of a mitochondrion from an active muscle cell would differ from the one shown. Give a reason for your answer.

Structural difference _____

_____   **1**

Reason _____

_____   **1**

Marks

3. The family tree shows the inheritance of a bone disorder.

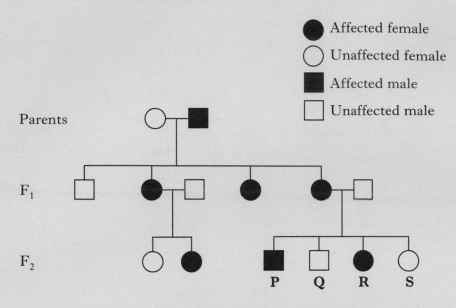

The disorder is caused by a dominant sex-linked allele (**B**).

(*a*) Using appropriate symbols, give the genotypes of individuals **P**, **Q**, **R** and **S**.

P——————— **Q**——————— **R**——————— **S**——————— 2

(*b*) (i) Explain why all the F$_1$ females in this family are affected.

_____

_____

_____ 1

(ii) Explain why only some of the F$_2$ females in this family are affected.

_____

_____

_____ 1

(*c*) Is the ratio of affected offspring to unaffected offspring in the F$_1$ generation as expected? Give a reason for your answer.

Yes/No ——————

Reason _____

_____ 1

Marks

4. The diagrams below show a disease-causing virus and one of the same type which has been weakened to make it less harmful.

Disease-causing virus   Weakened virus

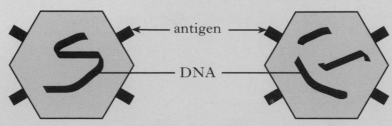

←—— antigen ——→

——— DNA ———

(*a*) A woman is vaccinated with the weakened form of the virus.

   (i) Explain why she does not develop the disease from the vaccination.

   _____

   _____   1

   (ii) What feature of the weakened virus results in her gaining immunity from the disease?

   _____

   _____   1

   (iii) Explain why this form of immunity is described as being both artificial and active.

   Artificial _____

   _____   1

   Active _____

   _____   1

(*b*) The table contains information about viruses.

   Tick (✓) the appropriate boxes to show characteristics which apply to all viruses.

Characteristic	*Tick* (✓)
Contains a nucleus	
Surrounded by a protein coat	
Can be seen under a light microscope	
Contains nucleic acid	
Can only reproduce inside other cells	

2

Marks

5. (*a*) The diagrams below contain information about three hormones involved in the control of milk production.

Placenta

Hormone X   Hormone Y

Pituitary Gland

Pituitary Gland

Hormone Z

| Mammary Glands
manufacture milk |
|:---:|

•••••••► inhibition

──────► stimulation

(i) Names hormones X, Y and Z.

X _____

Y _____

Z _____   2

(ii) Placental hormones inhibit the production of hormone Z by the pituitary gland. With reference to the diagrams, explain why milk production starts after birth.

_____

_____

_____   2

(*b*) (i) What name is given to the first milk produced by the mammary glands?

_____   1

(ii) State **one** difference in the content of this first milk compared with breast milk which is produced later.

_____   1

(*c*) Complete the following table which contains information about hormones produced by the pituitary gland.

Name of hormone	Target organ	Effect of hormone on target organ
ADH	kidney	
	testes	testosterone production
oxytocin		muscular contraction

2

Marks

6. The graph below shows the occurrence of high blood pressure in British men of different ages.

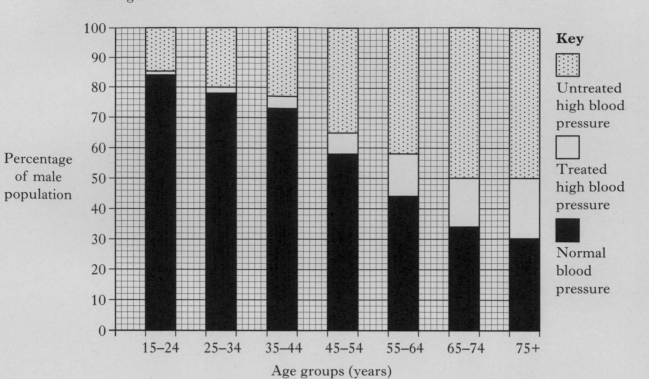

Percentage of male population *(y-axis)*

Age groups (years) *(x-axis)*

Key

Untreated high blood pressure

Treated high blood pressure

Normal blood pressure

(*a*) (i) What percentage of British men aged between 25 and 34 have high blood pressure?

_____   1

(ii) In men aged 55–64 who have high blood pressure, what is the percentage of treated to untreated individuals expressed as a simple ratio?

Space for working

_____ : _____   1

treated   untreated

(iii) Describe **one** trend shown by the graph and suggest an explanation for it.

Trend _____   1

Explanation _____

_____

_____   1

Marks

6. **(continued)**

(b) A blood pressure reading that is greater than 160/90 mmHg is regarded as being too high.

Why are blood pressure readings given as two figures?

_____

_____

_____

_____ 1

(c) Beta-blockers are drugs often used in the treatment of high blood pressure.

(i) Beta-blockers cause vasodilation. Explain how this lowers blood pressure.

_____

_____

_____

_____ 1

(ii) Beta-blockers also slow heart rate. Suggest which region of the heart is likely to be affected by beta-blockers.

_____ 1

[Turn over

Marks

7. The diagram below shows the liver, intestine and associated blood vessels.

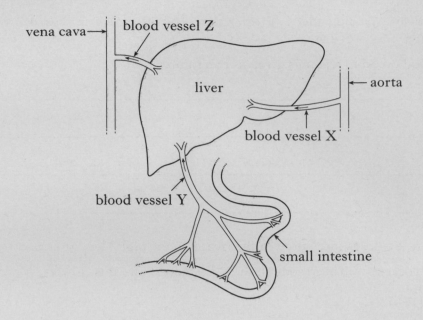

(a) (i) Identify blood vessels X, Y and Z.

X _____

Y _____

Z _____   **2**

(ii) Describe the differences in oxygen and carbon dioxide concentrations between blood vessel X and blood vessel Z.

_____

_____

_____   **1**

(b) (i) Glucose is absorbed into the blood stream from the small intestine.

Describe **two** ways in which the small intestine is designed to maximise glucose absorption.

1 _____

_____

2 _____

_____   **1**

DO NOT
WRITE IN
THIS
MARGIN

Marks

7. (*b*) (continued)

(ii) Describe **two** possible fates of the absorbed glucose when it reaches the liver.

1 _____

_____

2 _____

_____   1

(*c*) The liver metabolises a large number of substances.

(i) Name a substance excreted from the liver when red blood cells are broken down.

_____   1

(ii) What compounds are broken down in the liver to produce urea?

_____   1

[Turn over

Marks

8. The graphs below contain information about the regulation of blood sugar.

    Graph 1 shows how the concentration of glucose in the blood affects the concentration of insulin.

    Graph 2 shows how the concentration of insulin in the blood affects the rate of glucose uptake by the liver.

    Graph 1

    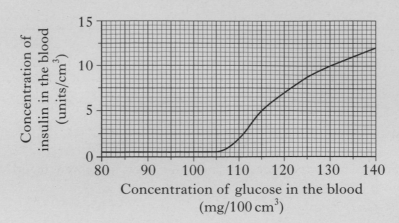

    Graph 2

    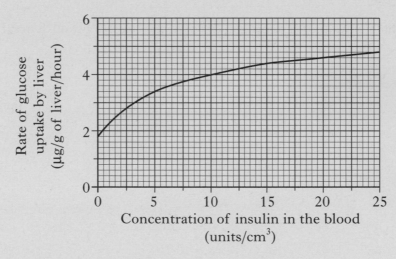

    (*a*) (i) From **Graph 1**, state the glucose concentration which triggers an increase in insulin production.

    _____   1

    (ii) Name the organ which produces insulin.

    _____   1

    (*b*) From **Graph 2**, calculate the percentage increase in the rate of glucose uptake by the liver when the concentration of insulin in the blood rises from 10 to 15 units/cm³.

    Space for calculation

    _____   1

    (*c*) From **Graphs 1** and **2**, state the rate of glucose uptake by the liver when the concentration of glucose in the blood is 130 mg/100 cm³.

    _____ μg/g of liver/hour   1

Marks

9. The diagram shows a section through part of the central nervous system.

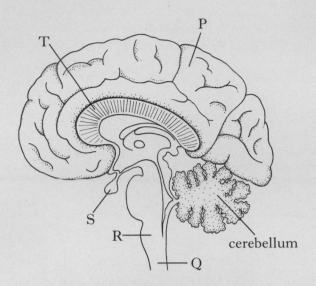

(*a*) The table contains information about three parts of the central nervous system. Complete the table to identify the parts and their functions.

Label	Name	Function
		Controls voluntary actions
T		Links left and right side of brain
	Spinal cord	

3

(*b*) Complete the following sentences by underlining one option from each pair of options shown in **bold**.

The parasympathetic nervous system is part of the **autonomic** / **somatic**

nervous system which originates in the **medulla** / **cerebellum**.

Parasympathetic nerves **speed up** / **slow down** heart rate.

1

(*c*) What structural feature of motor and sensory neurones speeds up the transmission of nerve impulses?

_____

_____

1

[Turn over

Marks

10. An investigation was carried out to find out how an infant's play was affected by the presence or absence of an adult. The infant was tested at three-month intervals using the following procedure.

1   The infant was placed in a room with an adult and some toys.
2   The infant was allowed to play with the toys for five minutes, then the adult left the room.
3   The infant was allowed to continue to play with the toys for another five minutes alone.

Playing time was measured by the number of seconds the infant spent playing per minute.

The graph shows the change in time spent playing, at each age, after the adult left the room.

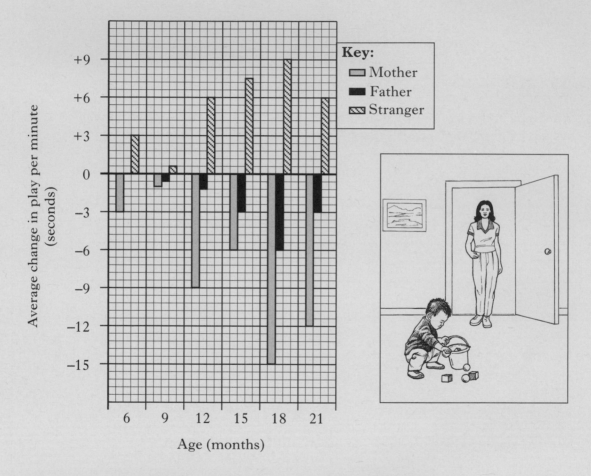

(a) At what ages does the departure of any adult have the **greatest** and **least** effect on the length of play time?

Greatest effect _____ months   Least effect _____ months   1

(b) When the child was 21 months old, what was the total increase in playing time, over the 5-minute period, when the stranger left the room?

_____ seconds   1

Marks

10. (continued)

(*c*)   (i)   Compare the effect of the departure of the mother with the departure of the father.

_____

_____

_____   **1**

     (ii)   Suggest a reason for this difference.

_____

_____

_____

_____   **1**

(*d*)   (i)   Compare the effect of the departure of the stranger with the departure of the parents.

_____

_____

_____   **1**

     (ii)   Suggest reasons for this difference.

_____

_____

_____

_____

_____

_____   **2**

(*e*)   How could the reliability of this investigation be improved?

_____   **1**

[Turn over

DO NO
WRITE
THIS
MARGI

Marks

11. An investigation was carried out on the effect of strobe lighting and loud noise on the ability of students to perform calculations.

Twenty students were divided into two equal groups, A and B. Each group was given 20 calculations to complete.

Group A sat in an evenly lit, quiet room.
Group B sat in a room where there was strobe lighting and loud noise.

The numbers of errors the students made, while doing the calculations, are shown in **Table 1**.

Table 1

Group A		Group B	
Student	Number of errors	Student	Number of errors
1	2	1	8
2	4	2	5
3	3	3	9
4	1	4	4
5	3	5	6
6	0	6	3
7	2	7	4
8	3	8	7
9	1	9	6
10	1	10	8

(a) By how many times has the average number of errors increased as a result of the distractions?

Space for calculation

_____ 1

(b) State **three** factors which would need to be kept constant during this investigation.

1 _____

2 _____

3 _____ 2

Marks

11. **(continued)**

(*c*) A third group of ten students carried out the investigation under the same conditions as group B, but were tested six times instead of only once. Each test comprised different calculations. The average percentage of errors is shown in **Table 2**.

Table 2

Trial	1	2	3	4	5	6
Average percentage error	34	30	24	20	20	19

(i) Construct a line graph to show the data in the table.

(Additional graph paper, if required, can be found on page 32.)

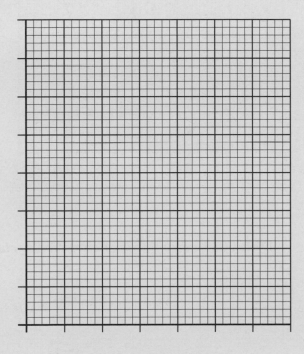

2

(ii) Suggest an explanation for the shape of the graph.

_____

_____

1

(*d*) How could the design of the investigation be altered to demonstrate the effect of social facilitation?

_____

_____

_____

1

12. **Graph A** shows how the average global temperature, between 1860 and 2000, varied from that in 1970.

Graph A

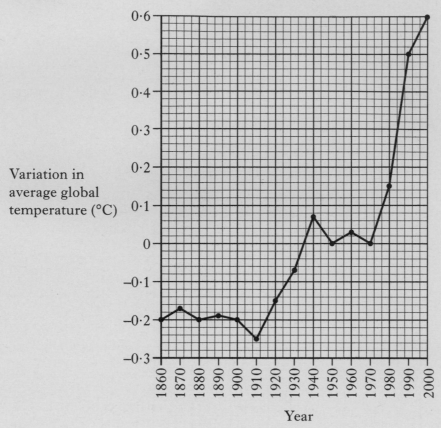

Variation in average global temperature (°C)

Year

Graph B shows the global fossil fuel consumption, between 1860 and 2000.

Graph B

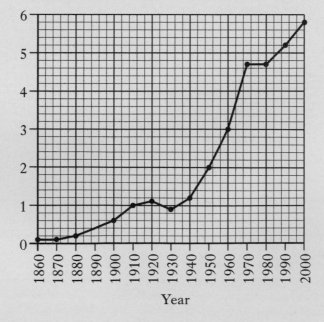

Global fossil fuel consumption (gigatones carbon/year)

Year

Marks

12.   (continued)

(a)   What was the increase in average global temperature between 1900 and 2000?

_____

1

(b)   State **two** reasons for the increased use of fossil fuels.

1 _____

2 _____

1

(c)   Discuss the extent to which the graphs support the theory that rising global temperatures are due to increasing use of fossil fuels.

Quote data from the graph in your answer.

_____

_____

_____

_____

_____

_____

_____

_____

2

(d)   Name a greenhouse gas other than carbon dioxide.

_____

1

[Turn over

DO NO
WRITE
THIS
MARGI

Marks

13. The photograph below shows the effect of deforestation on an area of tropical rainforest.

(*a*)   (i)   State **two** reasons why humans remove forest from the land.

1 _____

_____

2 _____

_____   2

(ii)   Deforestation can result in desertification. Explain how this can happen.

_____

_____

_____

_____

_____   2

(*b*)   Describe **one** other effect of deforestation on the local environment.

_____

_____

_____   1

DO NOT
WRITE IN
THIS
MARGIN

Marks

SECTION C

Both questions in this section should be attempted.

Note that each question contains a choice.

Questions 1 and 2 should be attempted on the blank pages which follow.

Supplementary sheets, if required, may be obtained from the invigilator.

Labelled diagrams may be used where appropriate.

1. Answer **either** A **or** B.

    A. Give an account of the function of a synapse under the following headings:

        (i) release of neurotransmitter; **3**

        (ii) action of neurotransmitter; **3**

        (iii) removal of neurotransmitter. **4**

        (10)

    OR

    B. Give an account of memory under the following headings:

        (i) encoding into short-term memory; **2**

        (ii) transfer from short-term to long-term memory; **6**

        (iii) retrieval from long-term memory. **2**

        (10)

In question 2, ONE mark is available for coherence and ONE mark is available for relevance.

2. Answer **either** A **or** B.

    A. Give an account of the causes and treatment of female infertility. **(10)**

    OR

    B. Give an account of how the structure of a red blood cell relates to its function. **(10)**

[END OF QUESTION PAPER]

DO NO
WRITE
THIS
MARG

SPACE FOR ANSWERS

SPACE FOR ANSWERS

DO NO
WRITE
THIS
MARG

SPACE FOR ANSWERS

SPACE FOR ANSWERS

DO NO
WRITE
THIS
MARG

SPACE FOR ANSWERS

DO NOT
WRITE IN
THIS
MARGIN

DO NO
WRITE
THI
MARG

SPACE FOR ANSWERS

ADDITIONAL GRAPH PAPER FOR QUESTION 11(*c*)(i)

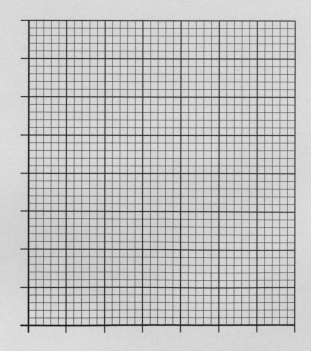

HIGHER

2008

[BLANK PAGE]

FOR OFFICIAL USE

Total for Sections B & C

X009/301

NATIONAL QUALIFICATIONS 2008

TUESDAY, 27 MAY 1.00 PM – 3.30 PM

HUMAN BIOLOGY HIGHER

Fill in these boxes and read what is printed below.

Full name of centre

Town

Forename(s)

Surname

Date of birth

Day Month Year

Scottish candidate number

Number of seat

SECTION A—Questions 1–30

Instructions for completion of Section A are given on page two.

For this section of the examination you must use an **HB pencil**.

SECTIONS B AND C

1   (a) All questions should be attempted.

   (b) It should be noted that in **Section C** questions 1 and 2 each contain a choice.

2   The questions may be answered in any order but all answers are to be written in the spaces provided in this answer book, **and must be written clearly and legibly in ink**.

3   Additional space for answers will be found at the end of the book. If further space is required, supplementary sheets may be obtained from the invigilator and should be inserted inside the **front** cover of this book.

4   The numbers of questions must be clearly inserted with any answers written in the additional space.

5   Rough work, if any should be necessary, should be written in this book and then scored through when the fair copy has been written. If further space is required a supplementary sheet for rough work may be obtained from the invigilator.

6   Before leaving the examination room you must give this book to the invigilator. If you do not, you may lose all the marks for this paper.

Read carefully

1   Check that the answer sheet provided is for **Human Biology Higher (Section A)**.

2   For this section of the examination you must use an **HB pencil**, and where necessary, an eraser.

3   Check that the answer sheet you have been given has **your name**, **date of birth**, **SCN** (Scottish Candidate Number) and **Centre Name** printed on it.

    Do not change any of these details.

4   If any of this information is wrong, tell the Invigilator immediately.

5   If this information is correct, **print** your name and seat number in the boxes provided.

6   The answer to each question is **either** A, B, C or D. Decide what your answer is, then, using your pencil, put a horizontal line in the space provided (see sample question below).

7   There is **only one correct** answer to each question.

8   Any rough working should be done on the question paper or the rough working sheet, **not** on your answer sheet.

9   At the end of the exam, put the **answer sheet for Section A inside the front cover of this answer book**.

Sample Question

The digestive enzyme pepsin is most active in the

A   stomach

B   mouth

C   duodenum

D   pancreas.

The correct answer is **A**—stomach. The answer **A** has been clearly marked in **pencil** with a horizontal line (see below).

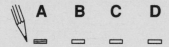

Changing an answer

If you decide to change your answer, carefully erase your first answer and, using your pencil, fill in the answer you want. The answer below has been changed to **D**.

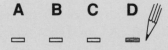

SECTION A

All questions in this section should be attempted.

Answers should be given on the separate answer sheet provided.

1. During which of the following chemical conversions is ATP produced?

   A   Amino acids ———→ protein

   B   Glucose ———→ pyruvic acid

   C   Haemoglobin ———→ oxyhaemoglobin

   D   Nucleotides ———→ mRNA

2. The following statements relate to respiration and the mitochondrion.

   1   Glycolysis takes place in the mitochondrion.

   2   The mitochondrion has two membranes.

   3   The rate of respiration is affected by temperature.

   Which of the above statements are correct?

   A   1 and 2

   B   1 and 3

   C   2 and 3

   D   All of them

3. The anaerobic breakdown of glucose splits from the aerobic pathway of respiration

   A   after the formation of pyruvic acid

   B   after the formation of acetyl CoA

   C   after the formation of citric acid

   D   at the start of glycolysis.

4. In respiration, the products of the cytochrome system are

   A   hydrogen and carbon dioxide

   B   water and ATP

   C   oxygen and ADP

   D   pyruvic acid and water.

5. The key below can be used to identify carbohydrates.

   1   Soluble.....................................2
        Insoluble..................................glycogen

   2   Benedict's test positive.............3
        Benedict's test negative............sucrose

   3   Barfoed's test positive 4
        Barfoed's test negative ....lactose

   4   Clinistix test positive ....glucose
        Clinistix test negative ....fructose

   Which line in the table of results below is **not** in agreement with the information contained in the key?

	Carbohydrate	Benedict's test	Barfoed's test	Clinistix test
A	lactose	positive	negative	not tested
B	glucose	positive	negative	positive
C	fructose	positive	positive	negative
D	sucrose	negative	not tested	not tested

6. Which of the following is an immune response?

   A   T-lymphocytes secreting antigens

   B   T-lymphocytes carrying out phagocytosis

   C   B-lymphocytes combining with foreign antigens

   D   B-lymphocytes producing antibodies

7. Phagocytes contain many lysosomes so that

   A   enzymes which destroy bacteria can be stored

   B   toxins from bacteria are neutralised

   C   antibodies can be released in response to antigens

   D   bacteria can be engulfed into the cytoplasm.

8. Which of the following is an example of active immunity?

   A   Antibody production following exposure to antigens

   B   Antibodies crossing the placenta from mother to fetus

   C   Antibodies passing from the mother's milk to a suckling baby

   D   Antibody extraction from one mammal to inject into another

9. The following steps occur during the replication of a virus.

   1   Alteration of host's cell metabolism

   2   Production of viral protein coats

   3   Replication of viral DNA

   In which sequence do these events occur?

   A   1 → 3 → 2

   B   1 → 2 → 3

   C   2 → 1 → 3

   D   3 → 1 → 2

10. The diagram below shows a stage in meiosis.

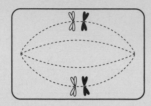

   Which of the following diagrams shows the next stage in meiosis?

A

B

C
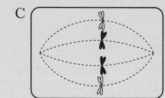

D
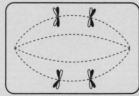

11. Cystic fibrosis is a genetic condition caused by an allele which is not sex-linked.

   A child is born with cystic fibrosis despite neither parent having the condition.

   The parents are going to have a second child. What is the chance this child will have cystic fibrosis?

   A   75%

   B   67%

   C   50%

   D   25%

12. A sex-linked condition in humans is caused by a recessive allele. What is the chance of an unaffected man and a carrier woman having an unaffected male child?

   A   1 in 1

   B   1 in 2

   C   1 in 3

   D   1 in 4

13. One function of the seminal vesicles is to

   A   produce testosterone

   B   allow sperm to mature

   C   store sperm temporarily

   D   produce nutrients for sperm.

14. Which fertility treatment would be appropriate for a woman with blocked uterine tubes?

   A   Provision of fertility drugs

   B   *In vitro* fertilisation

   C   Artificial insemination

   D   Calculation of fertile period

15. A 30 g serving of breakfast cereal contains 1·5 mg of iron. Only 25% of this iron is absorbed into the bloodstream.

   If a pregnant woman requires 6 mg of iron per day, how much cereal would she have to eat each day to meet this requirement?

   A   60 g

   B   120 g

   C   240 g

   D   480 g

16. Which of the following blood vessels carries oxygenated blood?

   A   Renal vein

   B   Hepatic vein

   C   Pulmonary vein

   D   Hepatic portal vein

17. In which of the following situations might a fetus be at risk from Rhesus antibodies produced by the mother?

	Father	Mother
A	Rhesus positive	Rhesus negative
B	Rhesus positive	Rhesus positive
C	Rhesus negative	Rhesus negative
D	Rhesus negative	Rhesus positive

18. The diagram below shows an ECG trace taken during exercise.

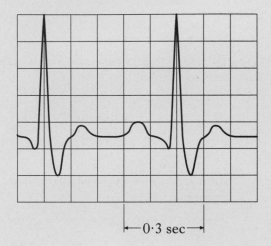

←—0·3 sec—→

The person's heart rate is

A   80 bpm

B   100 bpm

C   120 bpm

D   140 bpm.

19. The diagram below shows a section through the human heart.

What is the correct position of the pacemaker?

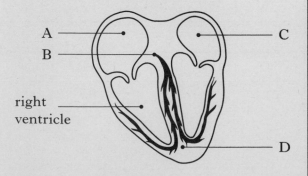

20. The vessel by which blood leaves the liver is the

A   renal vein

B   hepatic portal vein

C   renal artery

D   hepatic vein.

21. The graph below shows an individual's skin temperature and rate of sweat production over a period of 50 minutes.

Key
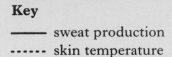
——— sweat production
------ skin temperature

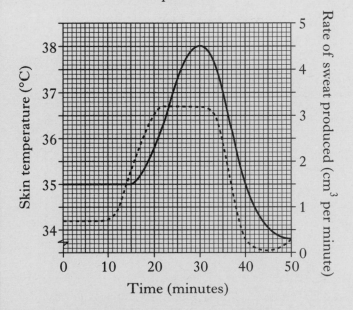

What is the skin temperature when the rate of sweat production is at a maximum?

A   3·2 °C

B   4·5 °C

C   36·7 °C

D   38·0 °C

[Turn over

22. The following diagram represents four neurones in a neural pathway.

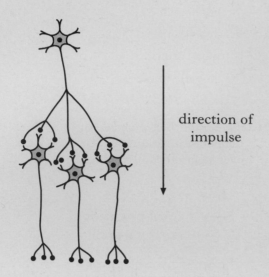

direction of impulse

Which line in the table describes the pathway correctly?

	Type of pathway	
A	motor	divergent
B	motor	convergent
C	sensory	divergent
D	sensory	convergent

23. Which of the following carries an impulse towards a nerve cell body?

A   Dendrite

B   Axon

C   Myelin

D   Myosin

24. Which of the following statements describes a neurotransmitter and its method of removal?

A   Adrenaline is removed by reabsorption.

B   Adrenaline is removed by enzyme degradation.

C   Noradrenaline is removed by enzyme degradation.

D   Noradrenaline is removed by reabsorption.

25. The diagram below illustrates the relationship between short and long-term memory.

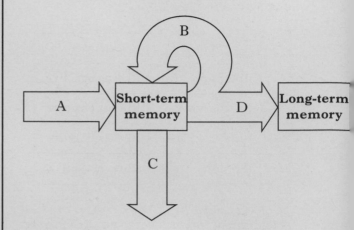

Which arrow represents the process of rehearsal?

26. The behavioural term *generalisation* is defined correctly as the ability to

A   make appropriate different responses to different but related stimuli

B   make the same appropriate response to different but related stimuli

C   submerge one's personal identity in the anonymity of a group

D   improve performance in competitive situations.

27. The table below contains information about the populations of four countries in the year 2000.

In which country did the population decrease?

Country	Number per 1000 inhabitants			
	Births	Deaths	Immigrants	Emigrants
A	9·3	10·1	1·0	0·1
B	9·7	10·3	1·3	0·4
C	10·1	9·9	0·2	0·5
D	10·8	10·5	0·1	0·3

28. The diagram below shows the number of people dying from different causes in a developing country. (Figures are in millions.)

Causes of death

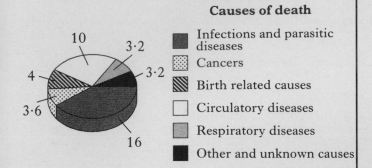

- ▨ Infections and parasitic diseases
- ▦ Cancers
- ▧ Birth related causes
- ☐ Circulatory diseases
- ▨ Respiratory diseases
- ■ Other and unknown causes

What percentage of deaths is due to birth related causes?

A   4%

B   8%

C   10%

D   11%

29. Which of the following processes is carried out by bacteria found in root nodules?

A   Denitrification

B   Nitrification

C   Nitrogen fixation

D   Deamination

30. Which of the following does **not** play a part in global warming?

A   The cutting down of forests

B   Methane production by cattle

C   The increase in use of motor vehicles

D   The increased use of fertilisers on farmland

Candidates are reminded that the answer sheet MUST be returned INSIDE the front cover of this answer booklet.

[Turn over for Section B

Marks

SECTION B

All questions in this section should be attempted.

All answers must be written clearly and legibly in ink.

1.   The diagram below illustrates the two main stages of protein synthesis.

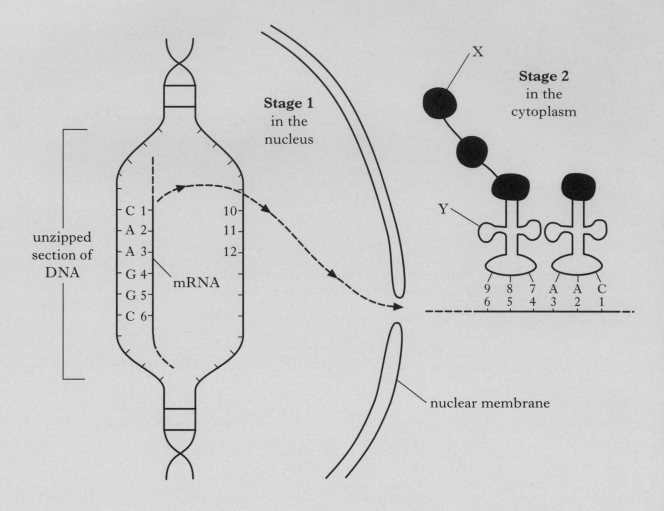

(a)   Describe **three** differences between DNA and mRNA.

1 _____

_____

2 _____

_____

3 _____

_____   2

Marks

1.   **(continued)**

(b)   Name bases 3, 8 and 11.

Base 3 _____

Base 8 _____

Base 11 _____   **2**

(c)   **Circle** a codon in the diagram opposite.   **1**

(d)   Where in the cytoplasm does stage 2 take place?

_____   **1**

(e)   Name molecules X and Y.

X _____   Y _____   **1**

(f)   The newly synthesised protein may be secreted from the cell.

(i)   Name the cell structure where the protein would be found just before it enters a secretory vesicle.

_____   **1**

(ii)   Describe what happens to the protein while it is in this cell structure.

_____   **1**

[Turn over

2. (*a*) The diagram below shows some of the functions of proteins in the cell membrane.

Marks

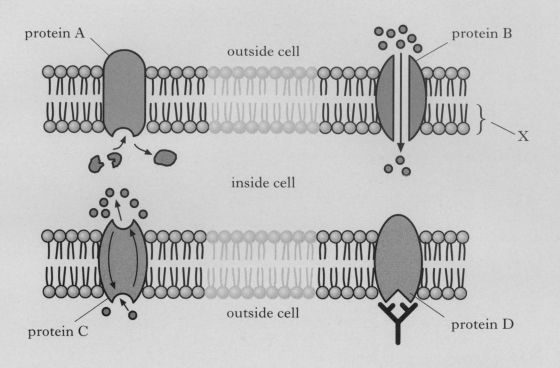

(i) Use the information from the diagram to complete the table below.

Protein	Function
	Transports molecules by diffusion
A	
D	
	Transports molecules by active transport

3

(ii) Identify molecule X and describe its function within the membrane.

Molecule X _____

Function _____

_____

2

(*b*) Describe what happens to the cell membrane during the process of endocytosis.

_____

_____

_____

2

Marks

3. The blood group of an individual is controlled by three alleles *A*, *B* and *O*.

Alleles *A* and *B* are co-dominant and completely dominant to allele *O*.

The diagram below shows the blood groups of three generations of a family.

Parents   Mother   Father
                       Group B   Group A

Children   Son 1   Son 2   Daughter ——— Husband
                Group A   Group O   Group O

Grandchildren   Grandson   Granddaughter
                                            Group B   Group A

(*a*) What is the blood group of the daughter?

_____   1

(*b*) State the genotypes of the grandchildren.

Grandson _____   Granddaughter _____   1

(*c*) How many of the three children are homozygous?

_____   1

(*d*) Explain the meaning of the term *co-dominant*.

_____

_____   1

(*e*) Only one of the sons can safely receive a blood transfusion from his brother.
Indicate whether this statement is true or false and explain your decision.

True/False _____

Explanation _____

_____

_____

_____   2

[Turn over

DO NO
WRITE
THIS
MARGI

4. The graphs below show the plasma concentrations of certain hormones throughout a woman's menstrual cycle.

Graph 1 shows the concentrations of FSH and LH.

Graph 2 shows the concentration of two other hormones, X and Y.

Graph 1

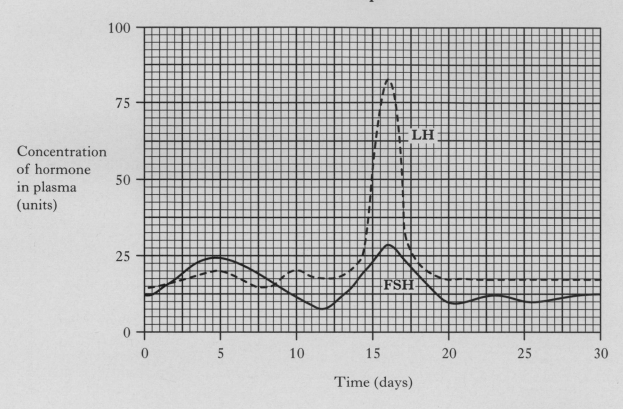

Graph 2

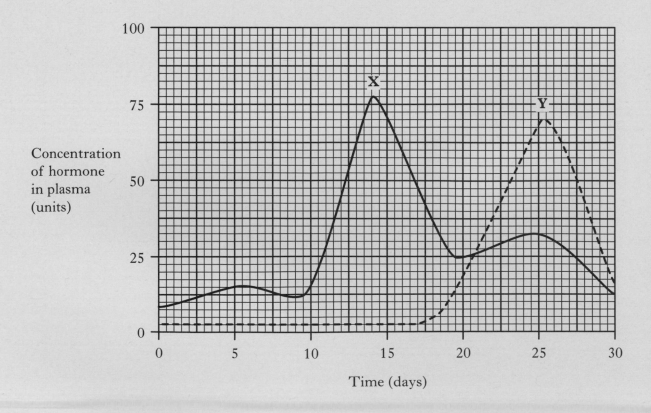

Marks

4.   **(continued)**

(a)   Where in the body are FSH and LH produced?

_____   1

(b)   Name hormones X and Y.

X   _____

Y   _____   1

(c)   What is the maximum concentration of hormone Y?

_____ units   1

(d)   On which day did ovulation occur? Give a reason for your answer.

Day   _____   1

Reason   _____

_____   1

(e)   During her next cycle, the woman became pregnant.

Describe any differences which would occur in the concentrations of FSH and hormone Y after day 25.

FSH   _____

_____   1

Hormone Y   _____

_____   1

[Turn over

Marks

5. (*a*) The table shows average quantities of substances filtered and excreted by the kidney per day.

Substance	Quantity filtered per day	Quantity excreted per day	Quantity reabsorbed per day
Water	$180 \, dm^3$	$1 \cdot 5 \, dm^3$	
Glucose	175 g	0 g	
Urea	48 g	31 g	
Protein	0 g	0 g	0 g

   (i) Complete the table by calculating the quantities reabsorbed per day for water, glucose and urea.

   1

   (ii) What percentage of water filtered by the kidney is reabsorbed?

   Space for calculation

   _____ %   **1**

   (iii) In which part of the kidney tubule is glucose reabsorbed?

   _____   **1**

(*b*) Nephrosis is a kidney condition in which glomeruli are damaged.

   As a result of nephrosis, the quantity of soluble proteins in the blood decreases and there is a build up of tissue fluid in the body.

   (i) Explain why damage to the glomeruli results in a decrease of soluble protein in the blood.

   _____

   _____   **1**

   (ii) Suggest a reason for the build-up of tissue fluid in the body.

   _____

   _____   **1**

Marks

6. The graph shows average blood pressure in different types of blood vessels.

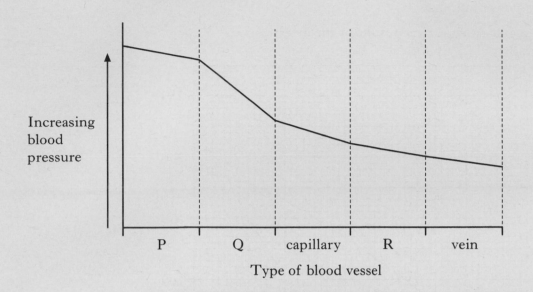

(*a*) Name the types of blood vessel represented by P, Q and R.

P _____

Q _____

R _____   2

(*b*) Blood pressure values fluctuate in vessel type P.

Explain the reason for this.

_____

_____   1

(*c*) Explain why there is a large drop in blood pressure in vessel type Q.

_____

_____

_____   1

(*d*) In the vena cava, blood pressure falls below atmospheric air pressure yet blood is still able to return to the heart.

Explain how the blood flow is maintained.

_____

_____

_____

_____   2

DO NC
WRITE
THIS
MARG

Marks

7. An investigation was carried out to find out how a cyclist's metabolism changed while he pedalled at increasing speed.

The cyclist's heart rate, fat and carbohydrate consumption were measured at different power outputs.

The graph below shows the results of the investigation.

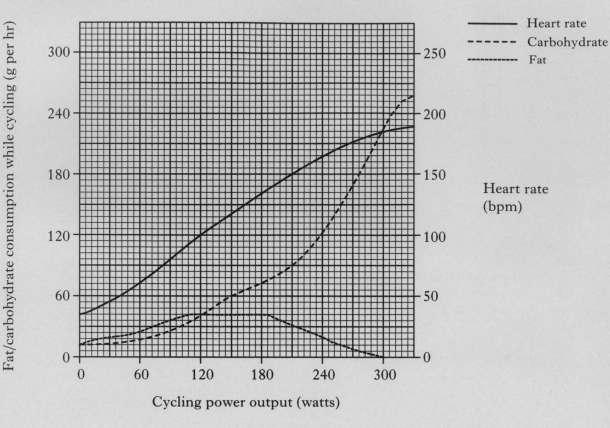

(a) What is the heart rate of the cyclist when his power output is 90 watts?

_____ bpm

1

(b) What evidence is there from the graph that the cyclist is very fit?

_____

_____

1

(c) Compare the consumption of fat and carbohydrate as cycling power increases. Quote data from the graph in your answer.

_____

_____

_____

_____

_____

_____

3

Marks

7.   (continued)

(d)   (i)   Cyclists often use heart-rate monitors in training. A cyclist wishes to maintain his fat consumption at its maximum and, at the same time, limit his carbohydrate consumption.

At what heart rate should he cycle?

_____ bpm

1

(ii)   Suggest why it is good practice in a long distance cycling race to maximise fat consumption and minimise carbohydrate consumption.

_____

_____

_____

1

(e)   The cyclist raced for 4 hours at a power output of 210 watts. During that time he consumed 100 g of carbohydrate in a liquid drink. Assuming he started with a carbohydrate store of 500 g, how much carbohydrate would he be left with at the end of the race?

Space for calculation

_____ g   1

(f)   (i)   Glycogen is a major source of carbohydrate. Where is glycogen stored in the body?

_____   1

(ii)   Name a hormone which promotes the conversion of glycogen to glucose.

_____   1

(iii)   What substance is used as a source of energy after glycogen and fat stores have been used up?

_____   1

[Turn over

Marks

8. The diagrams below show two possible ways of classifying the nervous system.

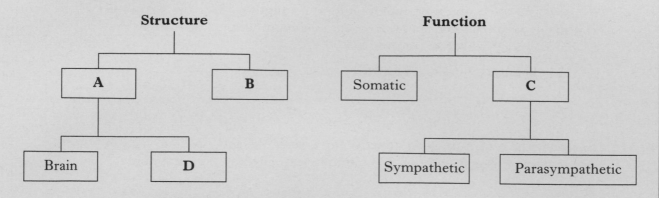

(a) (i) Identify A to D.

A _____

B _____

C _____

D _____   **2**

(ii) Describe **one** function of the somatic nervous system.

_____

_____

_____   **1**

(b) The brain contains two cerebral hemispheres.

(i) Name the structure which links these two hemispheres.

_____   **1**

(ii) The surfaces of the hemispheres are heavily folded to provide a large surface area.

Explain the significance of this feature.

_____

_____

_____   **1**

Marks

8. (continued)

(c) The diagram below shows some of the nerve connections between the brain and three parts of the body.

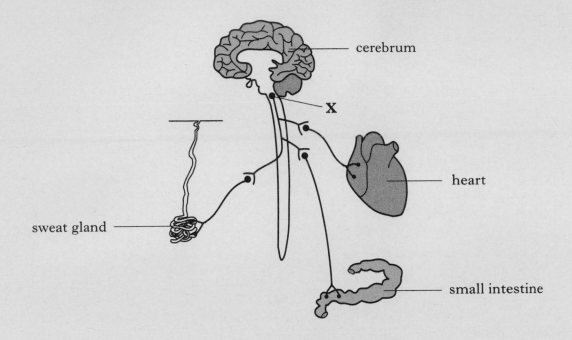

(i) Identify the part of the brain labelled **X**.

_____

1

(ii) The sympathetic and parasympathetic systems are often described as antagonistic to one another.

Explain the meaning of *antagonistic*.

_____

_____

1

(iii) Complete the table to show the effect of sympathetic stimulation on the heart, sweat glands and small intestine.

Part of body	Sympathetic effect
Heart	
Sweat glands	
Small intestine	

2

[Turn over

Marks

9. The diagram shows how a non-biodegradable insecticide passes through a food chain in a Scottish fresh-water loch.

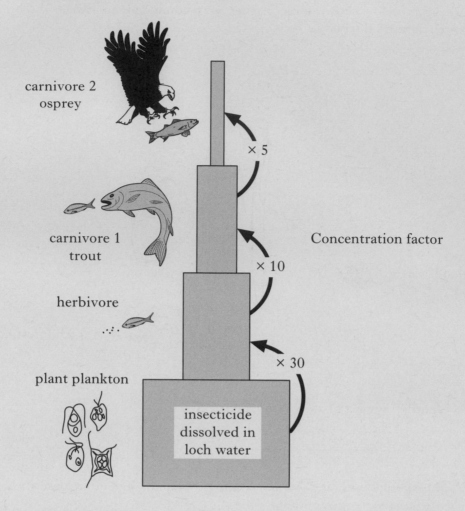

(a) Describe **one** way in which the insecticide could get into the loch water.

_____

_____

_____   1

(b) (i) The diagram shows the number of times the insecticide becomes concentrated at each stage of the food chain.

If the concentration of insecticide in the plant plankton is 0·025 ppm what would be the expected concentration in the osprey?

Space for calculation

_____ ppm   1

DO NOT
WRITE IN
THIS
MARGIN

Marks

9. **(b) (continued)**

(ii) Explain why insecticide becomes more concentrated in carnivores at the top of the food chain.

_____

_____

_____

_____ **2**

(c) DDT is an insecticide which breaks down slowly at a rate of 50% every fifteen years. Calculate how long it would take for 100 kg of DDT to break down to less than 1 kg.

Space for calculation

_____ years **1**

(d) Insecticides are chemicals used extensively in agriculture.

Name **two** other types of chemical used to treat crops and explain why they are used.

Chemical 1 _____

Use _____

_____ **1**

Chemical 2 _____

Use _____

_____ **1**

(e) Some insecticides work by disrupting enzyme-catalysed pathways.

What term is used to describe their action on enzymes?

_____ **1**

[Turn over

DO NO
WRITE
THIS
MARG

10. An experiment was carried out to investigate the effect of pH on the activity of the enzyme pepsin.

Marks

Six beakers were filled with pepsin solution and the pH adjusted in each beaker to give a range from pH 1 to pH 9. Six glass tubes were filled with egg albumen and boiled in water to set the egg white. The starting lengths of the egg white were measured and recorded in the table below.

The glass tubes were placed in the pepsin solution for a number of hours to allow digestion of the egg white. The lengths of egg white left in each tube at the end of the investigation are shown in the diagram below.

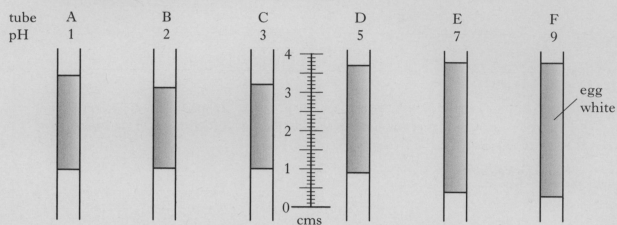

Tube	pH of pepsin solution	Length of egg white at start (mm)	Length of egg white at finish (mm)	Percentage decrease in length (%)
A	1	36	24	33
B	2	35	20	43
C	3	36		
D	5	34		
E	7	36	34	6
F	9	35	35	0

(*a*)　(i)　Complete the table above by measuring and recording the final lengths of egg white in tubes C and D.

1

　　　(ii)　Calculate the percentage decrease in length of egg white in tubes C and D and complete the table.

1

(*b*)　Draw a line graph to show the relationship between pH and percentage decrease in length of egg white.

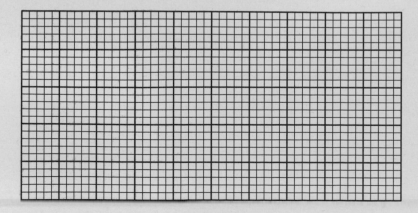

(Additional graph paper, if required, can be found on page 28.)

2

10. **(continued)**

Marks

(c)   (i)   What conclusion can be drawn from the results of this experiment?

_____

_____ **1**

(ii)   Predict the percentage decrease in length of egg white in a pepsin solution of pH 12.

_____ **1**

(iii)   Why was it necessary to calculate a *percentage* decrease?

_____

_____ **1**

(iv)   Describe a suitable control for tube **A** in this investigation.

_____

_____ **1**

(v)   State **three** variables which would have to be kept constant throughout this investigation.

1 _____

2 _____

3 _____ **2**

(vi)   Describe **one** way in which the results could be made more reliable.

_____

_____ **1**

(d)   Pepsin is produced in an inactive form by cells lining the stomach.

Why is it important that pepsin is inactive when it is produced?

_____

_____

_____ **1**

[Turn over

DO N(
WRITI
THI
MAR(

Marks

SECTION C

Both questions in this section should be attempted.

Note that each question contains a choice.

Questions 1 and 2 should be attempted on the blank pages which follow.

Supplementary sheets, if required, may be obtained from the invigilator.

Labelled diagrams may be used where appropriate.

1. Answer **either** A **or** B.

   A. Give an account of temperature regulation in cold conditions under the following headings:

      (i) voluntary responses;   3

      (ii) involuntary responses;   5

      (iii) hypothermia.   2

                                                              (10)

   OR

   B. Give an account of the development of boys at puberty under the following headings:

      (i) physical changes;   3

      (ii) hormonal changes.   7

                                                              (10)

In question 2, ONE mark is available for coherence and ONE mark is available for relevance.

2. Answer **either** A **or** B.

   A. Discuss how the impact of disease on the human population can be reduced.   **(10)**

   OR

   B. Describe the factors which influence the development of behaviour.   **(10)**

[END OF QUESTION PAPER]

DO NOT
WRITE IN
THIS
MARGIN

SPACE FOR ANSWERS

DO N
WRITE
THI
MARC

SPACE FOR ANSWERS

DO NOT
WRITE IN
THIS
MARGIN

SPACE FOR ANSWERS

DO NC
WRITE
THIS
MARG

SPACE FOR ANSWERS

ADDITIONAL GRAPH PAPER FOR QUESTION 10(*b*)

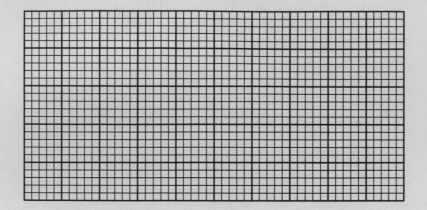

[BLANK PAGE]

FOR OFFICIAL USE

Total for
Sections B & C

X009/301

NATIONAL
QUALIFICATIONS
2009

THURSDAY, 28 MAY
1.00 PM – 3.30 PM

HUMAN BIOLOGY
HIGHER

Fill in these boxes and read what is printed below.

Full name of centre

Town

Forename(s)

Surname

Date of birth

Day   Month   Year   Scottish candidate number   Number of seat

SECTION A—Questions 1–30

Instructions for completion of Section A are given on page two.

For this section of the examination you must use an **HB pencil**.

SECTIONS B AND C

1   (a)   All questions should be attempted.

    (b)   It should be noted that in **Section C** questions 1 and 2 each contain a choice.

2   The questions may be answered in any order but all answers are to be written in the spaces provided in this answer book, **and must be written clearly and legibly in ink**.

3   Additional space for answers will be found at the end of the book. If further space is required, supplementary sheets may be obtained from the invigilator and should be inserted inside the **front** cover of this book.

4   The numbers of questions must be clearly inserted with any answers written in the additional space.

5   Rough work, if any should be necessary, should be written in this book and then scored through when the fair copy has been written. If further space is required a supplementary sheet for rough work may be obtained from the invigilator.

6   Before leaving the examination room you must give this book to the invigilator. If you do not, you may lose all the marks for this paper.

Read carefully

1   Check that the answer sheet provided is for **Human Biology Higher (Section A)**.

2   For this section of the examination you must use an **HB pencil**, and where necessary, an eraser.

3   Check that the answer sheet you have been given has **your name**, **date of birth**, **SCN** (Scottish Candidate Number) and **Centre Name** printed on it.

Do not change any of these details.

4   If any of this information is wrong, tell the Invigilator immediately.

5   If this information is correct, **print** your name and seat number in the boxes provided.

6   The answer to each question is **either** A, B, C or D. Decide what your answer is, then, using your pencil, put a horizontal line in the space provided (see sample question below).

7   There is **only one correct** answer to each question.

8   Any rough working should be done on the question paper or the rough working sheet, **not** on your answer sheet.

9   At the end of the exam, put the **answer sheet for Section A inside the front cover of this answer book**.

Sample Question

The digestive enzyme pepsin is most active in the

A   stomach

B   mouth

C   duodenum

D   pancreas.

The correct answer is **A**—stomach. The answer **A** has been clearly marked in **pencil** with a horizontal line (see below).

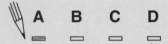

Changing an answer

If you decide to change your answer, carefully erase your first answer and, using your pencil, fill in the answer you want. The answer below has been changed to **D**.

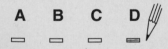

SECTION A

All questions in this section should be attempted.

Answers should be given on the separate answer sheet provided.

1. Which of the following often act as a co-enzyme?

   A   Lipids

   B   Polysaccharides

   C   Hormones

   D   Vitamins

2. The table below refers to the mass of DNA in certain human body cells.

Cell type	Mass of DNA in cell ($\times 10^{-12}$ g)
liver	6·6
lung	6·6
P	3·3
Q	0·0

   Which of the following is most likely to identify correctly cell types P and Q?

	P	Q
A	kidney cell	sperm cell
B	sperm cell	mature red blood cell
C	mature red blood cell	sperm cell
D	nerve cell	mature red blood cell

3. The diagram below shows energy transfer within a cell.

   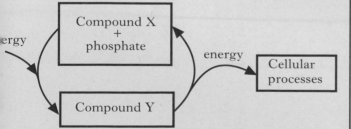

   Which line in the table below identifies correctly compounds X and Y?

	X	Y
A	glucose	ATP
B	glucose	ADP
C	ADP	ATP
D	ATP	glucose

4. The following chart shows stages in the complete breakdown of glucose in aerobic respiration.

   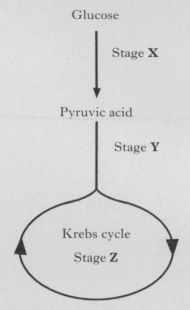

   At which stage or stages is hydrogen released to be picked up by hydrogen acceptors?

   A   Stages X, Y and Z

   B   Stages X and Y only

   C   Stages Y and Z only

   D   Stage Z only

5. The cell organelle shown below is magnified ten thousand times.

   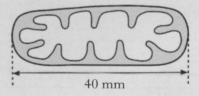

   What is the actual size of the organelle?

   A   0·04 μm

   B   0·4 μm

   C   4 μm

   D   40 μm

6. Lysosomes are abundant in

   A   enzyme secreting cells

   B   muscle cells

   C   cells involved in protein synthesis

   D   phagocytic cells.

7. The family tree below shows the transmission of the Rhesus D-antigen through three generations of a family. The allele coding for the presence of the Rhesus D-antigen is dominant and autosomal.

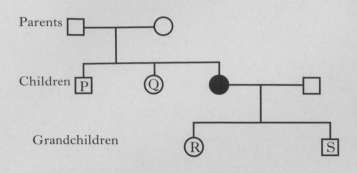

Parents

Children   P   Q

Grandchildren   R   S

☐   Rhesus positive male

■   Rhesus negative male

○   Rhesus positive female

●   Rhesus negative female

Which of the children and grandchildren in the family tree must be heterozygous?

A   P, Q, R and S

B   P and Q only

C   R and S only

D   Q and R only

8. The table below shows some genotypes and phenotypes associated with a form of anaemia.

Genotype	Phenotype
AA	Unaffected
AS	Sickle cell trait
SS	Acute sickle cell anaemia

An unaffected person and someone with sickle cell trait have a child together.

What are the chances of the child having acute sickle cell anaemia?

A   none

B   1 in 4

C   1 in 2

D   1 in 1

9. The graph below shows changes which occur in the masses of protein, fat and carbohydrate in a person's body during seven weeks without food.

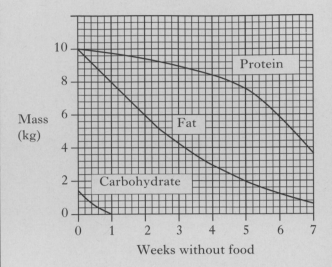

The person's starting weight was 60 kg. Predict their weight after two weeks without food.

A   57 kg

B   54 kg

C   50 kg

D   43 kg

10. The diagram below shows a section through seminiferous tubules in a testis.

Which cell produces testosterone?

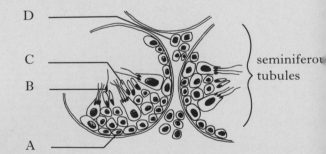

11. The diagram below represents part of the mechanism which controls ovulation.

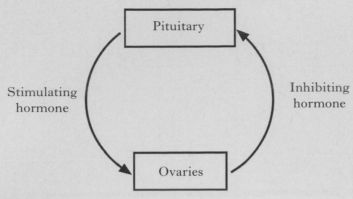

The hormones indicated above are

	Stimulating hormone	*Inhibiting hormone*
A	FSH	oestrogen
B	progesterone	FSH
C	oestrogen	LH
D	LH	testosterone

12. On which day in the following menstrual cycle could fertilisation occur?

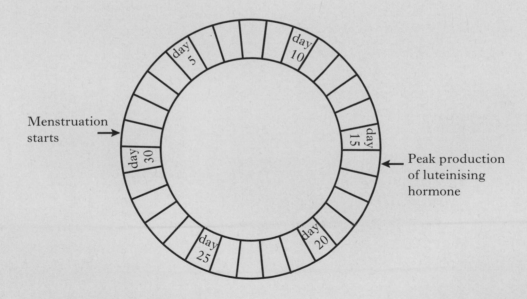

A   Day 30

B   Day 17

C   Day 14

D   Day 2

[Turn over

13. The diagram below shows blood vessels associated with the liver. The arrows show the direction of blood flow.

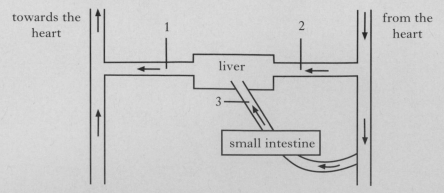

Which of the following correctly identifies the blood vessels.

	1	2	3
A	hepatic artery	hepatic vein	hepatic portal vein
B	hepatic vein	hepatic portal vein	hepatic artery
C	hepatic vein	hepatic artery	hepatic portal vein
D	hepatic artery	hepatic portal vein	hepatic vein

14. The relatively high urea concentration in the hepatic vein is a result of

   A reabsorption of amino acids in the kidney

   B conversion of glycogen to glucose in the liver

   C deamination of amino acids in the liver

   D excretion of amino acids in the kidney.

15. A person produces 0·75 litres of urine in 24 hours. The urine contains 18 g of urea.

   What is the concentration of urea in the urine?

   A $1·0\,g/100\,cm^3$

   B $1·8\,g/100\,cm^3$

   C $2·4\,g/litre$

   D $2·4\,g/100\,cm^3$

16. The diagram below represents a part of the circulatory system of the skin.

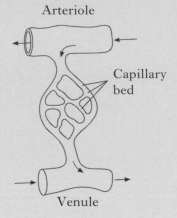

Which line in the table below correctly identifies changes which would take place in the blood as it flows from arteriole to venule?

	Concentration of	
	glucose	CO$_2$
A	increase	decrease
B	decrease	decrease
C	increase	increase
D	decrease	increase

17. A man was asked to breathe steadily at rest, then to breathe in and out as deeply as possible and finally to breathe steadily when exercising.

A trace of his lung capacity during this activity is shown.

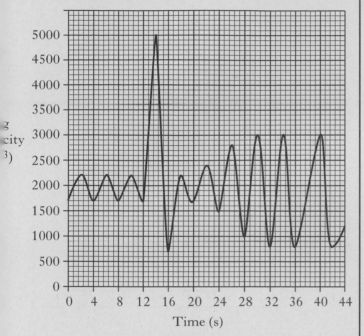

His volume of one breath at rest is

A $500\,cm^3$

B $2200\,cm^3$

C $4300\,cm^3$

D $5000\,cm^3$.

18. Which of the following is **not** a function of the lymphatic system?

A   It returns excess tissue fluid to the blood.

B   It causes the clotting of blood at wounds.

C   It destroys bacteria.

D   It transports fat from the small intestine.

19. When there is a decrease in the water concentration of the blood, which of the following series of events occur during the negative feedback response of the body?

	Concentration of ADH	Permeability of kidney tubules	Volume of urine
A	increases	increases	increases
B	decreases	decreases	increases
C	increases	increases	decreases
D	decreases	increases	decreases

20. Which of the following shows the correct responses to changes in blood sugar concentration?

	Sugar concentration in blood	Glucagon secretion	Insulin secretion	Glycogen stored in liver
A	increases	decreases	increases	increases
B	increases	decreases	increases	decreases
C	decreases	increases	decreases	increases
D	decreases	decreases	increases	decreases

[Turn over

21. High levels of blood glucose can cause clouding of the lens in the human eye. Concentrations above 5·5 mM are believed to put the individual at a high risk of lens damage.

In an investigation, subjects of different ages each drank a glucose solution. The concentration of glucose in their blood was then monitored over a number of hours. The results are shown in the graph below.

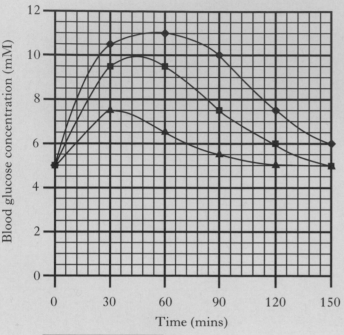

70 year olds

50 year olds

20 year olds

For how long during the investigation did 20 year olds remain above the high risk blood glucose concentration?

A   84 mins

B   90 mins

C   120 mins

D   148 mins

22. Which of the following parts of the brain is important in transferring information between the two cerebral hemispheres?

A   Hypothalamus

B   Corpus callosum

C   Cerebellum

D   Medulla oblongata

23. Which parts of the body are controlled by the largest motor area of the cerebrum?

A   Hands and lips

B   Feet and hands

C   Arms and hands

D   Legs and arms

24. The diagram below shows the ages in months at which children are able to walk unaided. The left end of the bar shows the age at which 25% of infants can walk unaided. The right end of the bar shows the age at which 90% of infants can walk unaided. The vertical bar shows the age at which 50% of infants can walk unaided.

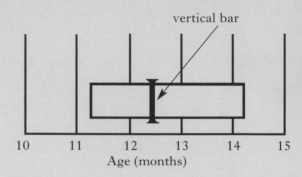

If sixteen infants, aged twelve months, were tested, how many would be expected to walk unaided?

A   4

B   7

C   9

D   12

25. Which of the following best describes memory span?

A   The total memory capacity of the brain

B   The time taken to learn a piece of information

C   The storage capacity of the short-term memory

D   The capacity to store information in long-term memory

26. The graph below shows the results of a survey carried out on members of an athletic club who ran an 800 m course under different conditions.

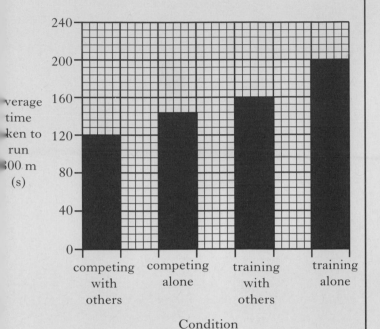

What is the percentage improvement in the time taken to run 800 m between those athletes training on their own and those training with others?

A   40%

B   25%

C   24%

D   20%

27. Which of the following processes reduces atmospheric carbon dioxide concentrations?

A   Decomposition

B   Nitrogen fixation

C   Respiration

D   Photosynthesis

28. Which of the following is a major source of methane?

A   Motor vehicles

B   Aerosols

C   Cattle

D   Nitrate fertilisers

29. The diagram below shows a nitrogen cycle associated with soil.

Which arrow indicates the activity of denitrifying bacteria?

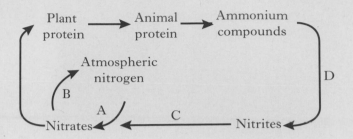

30. The age structure for four different human populations is represented in the diagrams below. The bars indicate the relative numbers in each age group.

Which diagram shows the population with the greatest potential for growth in the next forty years?

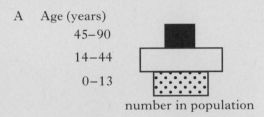

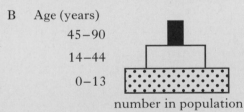

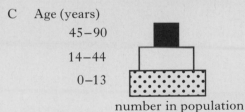

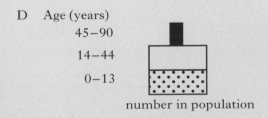

Candidates are reminded that the answer sheet MUST be returned INSIDE the front cover of this answer booklet.

Marks

SECTION B

All questions in this section should be attempted.

All answers must be written clearly and legibly in ink.

1. (a) The diagram below shows a section of a messenger RNA (mRNA) molecule.

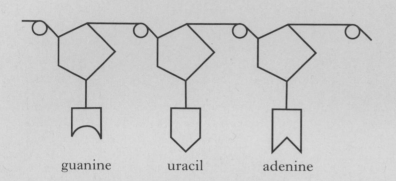

guanine   uracil   adenine

phosphate = ◯

sugar =

   (i) Name the sugar that is present in mRNA.

     _____

1

   (ii) Which base found in mRNA is **not** shown in the diagram?

     _____

1

   (iii) Name **two** parts of a cell where mRNA is found.

     1 _____

     2 _____

1

(b) DNA templates are used to produce mRNA molecules.

   (i) Insert the names of the DNA bases which pair with the RNA bases shown in the table below.

DNA base	RNA base
	adenine
	uracil
	guanine

1

Marks

1. (*b*) (continued)

(ii) Apart from free RNA nucleotides and a DNA template, name **one** other molecule that is essential for mRNA synthesis.

_____

1

(iii) Describe the part played by an mRNA molecule in the manufacture of a cell protein.

_____

_____

_____

_____

3

[Turn over

Marks

2.   Many inherited disorders are caused by inborn errors of metabolism.

(*a*)   (i)   What causes disorders that lead to inborn errors of metabolism?

_____   1

(ii)   How do these inherited disorders affect metabolic pathways?

_____

_____   1

(*b*)   Phenylketonuria (PKU) is an example of an inherited disorder.

One metabolic pathway affected by PKU is shown below.

$$\text{phenylalanine} \xrightarrow{\ enzyme\ 1\ } \text{tyrosine} \xrightarrow{\ enzyme\ 2\ } \text{intermediate compounds} \xrightarrow{\ enzyme\ 3\ } \text{noradrenaline}$$

(i)   Describe how PKU affects the metabolic pathway shown above.

_____

_____   1

(ii)   With reference to the metabolic pathway shown, explain why PKU affects the nervous system.

_____

_____

_____

_____   2

(*c*)   What term describes the testing of newborn babies for inherited disorders such as PKU?

_____   1

Marks

3. (*a*) The MN blood group system is determined by two alleles, M and N. Each of these alleles controls the production of a different antigen on the cell membrane of red blood cells.

M and N are co-dominant.

(i) Two parents, who are heterozygous for this blood group, have a son. Complete the Punnett square below to show the parental gametes and the possible genotypes of their son.

Parents   Mother   Father
Genotype   MN   ×   MN

		mother's gametes	
father's gametes			MN

1

(ii) The son has a different genotype to either of his parents. What are the chances of this happening?

Space for calculation

_____ %   1

(iii) Describe how the son's phenotype differs from his parents.

_____

_____   1

(*b*) The immune system recognises antigens on the cell membrane as self or non-self.

What term describes

(i) an immune reaction to self antigens?

_____   1

(ii) an over-reaction to a normally harmless non-self antigen?

_____   1

[Turn over

Marks

4. Hydrogen peroxide is a toxic chemical which is produced during metabolism. Catalase is an enzyme which breaks down hydrogen peroxide as shown below.

$$\text{hydrogen peroxide} \xrightarrow{\text{catalase}} \text{water} + \text{oxygen}$$

Experiments were carried out to investigate how changing the concentration of catalase affects the rate of hydrogen peroxide breakdown.

Filter paper discs were soaked in catalase solutions of different concentration. Each disc was then added to a beaker of hydrogen peroxide solution as shown in **Figure 1**.

The disc sank to the bottom of the beaker before rising back up to the surface. The time taken for each disc to rise to the surface was used to measure the reaction rates.

The results of the investigation are shown in **Table 1**.

Figure 1

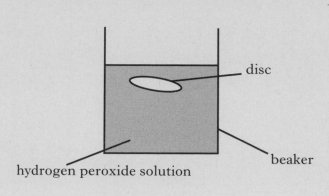

hydrogen peroxide solution

Table 1

catalase concentration (%)	average time for ten discs to rise (s)
0·125	9·8
0·25	6·9
0·5	5·0
1·0	3·8
2·0	3·8

(a) Explain why the filter paper discs rose to the surface of the hydrogen peroxide solution.

_____

_____   1

(b) Name **three** variables which should be controlled during this investigation.

1 _____

2 _____

3 _____   2

(c) What feature of **this** investigation makes the results more reliable?

_____

_____   1

Marks

4.   **(continued)**

(*d*) It was suggested that the filter paper was reacting with the hydrogen peroxide. How could this be tested using the same procedure?

_____

_____   1

(*e*)   (i) Plot a line graph to illustrate the results of the investigation. (Additional graph paper, if required, can be found on *Page thirty-two*.)

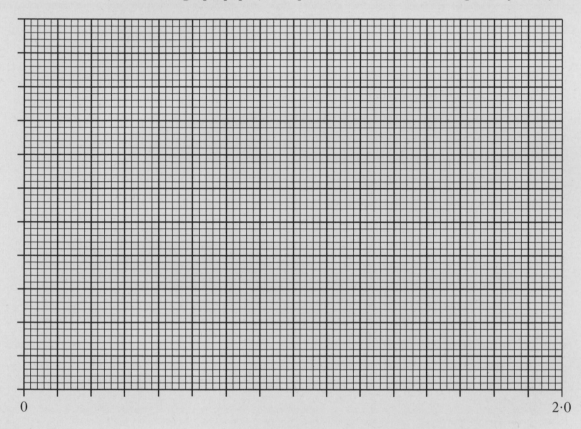

0   2·0   **2**

(ii) State **two** conclusions which can be drawn from these results.

1 _____

_____

2 _____

_____   **2**

(*f*) Explain why the addition of an inhibitor would slow down the rate of this reaction.

_____

_____   **1**

Marks

5. The diagram shows part of the reproductive system of a woman in early pregnancy.

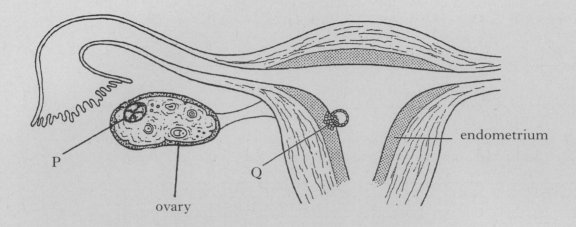

(*a*) Place an **X** on the diagram to show where fertilisation occurred.   1

(*b*) Structure P produces progesterone at this stage in pregnancy.

   (i) Name structure P.

      _____   1

   (ii) State **one** function of progesterone during early pregnancy.

      _____

      _____   1

(*c*) Structure Q will develop into the placenta.

Name the processes involved in the transfer of oxygen, glucose and antibodies across the placenta.

Oxygen _____

Glucose _____

Antibodies _____   2

(*d*) In the early stages of pregnancy the cells of the embryo are starting to differentiate.

Describe what happens during differentiation.

_____

_____   1

Marks

5. (continued)

(*e*) Name a stage of embryo development that comes between fertilisation and differentiation.

_____

1

(*f*) A woman gives birth to monozygotic twins.

State whether monozygotic twins are identical or non-identical and give a reason for your answer.

Monozygotic twins _____

Reason _____

_____

_____

1

[Turn over

DO N
WRIT
TH
MAR

6. The diagram below shows stages in the life history of a red blood cell.

Marks

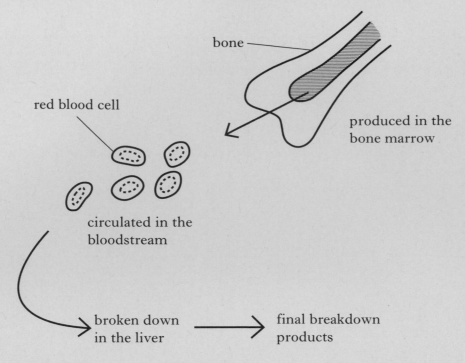

bone

produced in the
bone marrow

red blood cell

circulated in the
bloodstream

broken down
in the liver

final breakdown
products

(a) Vitamin B_{12} and iron are both used in the production of red blood cells.

(i) What substance is needed for the absorption of Vitamin B_{12} from the gut?

_____

1

(ii) Which molecule requires iron for its production?

_____

1

(b) On average, how long do red blood cells remain in circulation?

_____

1

(c) At any given time there are 5·5 million red blood cells in 1 millilitre of human blood.

Calculate how many red blood cells will be in the circulation of an individual who has a total blood volume of 5 litres.

Space for calculation

_____ million   1

Marks

6.   **(continued)**

(*d*)   Explain how the structure of a red blood cell

(i)   makes it very efficient at absorbing oxygen.

_____

_____   1

(ii)   allows it to pass through capillaries.

_____

_____   1

(*e*)   Apart from the liver, name a body site where red blood cells are broken down.

_____   1

(*f*)   One of the final products of the breakdown of red blood cells is bile.

(i)   Where is bile stored in the body?

_____   1

(ii)   Explain the importance of bile salts in the digestion of lipids.

_____

_____

_____

_____

_____   2

[Turn over

7. Oxygen consumption is often used to measure the intensity of exercise.

VO_{2max} is the maximum rate at which someone can take up and use oxygen.

Graph 1 shows the VO_{2max} of office workers, and various professional sportsmen and sportswomen.

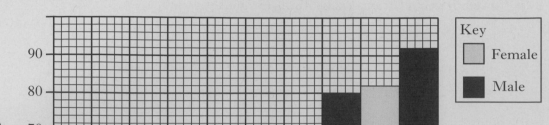

(a) (i) What is the difference between the VO_{2max} of a male cross-country skier and a male office worker?

Space for calculation

_____ **1**

(ii) Cross-country skiing is a very energy demanding sport.

What is the advantage to a cross-country skier of having a high VO_{2max}?

_____

_____ **1**

(b) Calculate the oxygen uptake, during a three minute race, of a female rower who weighs 85 kg. Assume that she has maximum oxygen uptake throughout the race.

Space for calculation

_____ litres **1**

(c) The graph shows that, on average, men have higher maximum oxygen uptakes than women.

Suggest a reason for this difference.

_____

_____ **1**

OFFICIAL SQA PAST PAPERS   119   HIGHER HUMAN BIOLOGY 2009

DO NOT
WRITE IN
THIS
MARGIN

Marks

7.   (continued)

Tests which determine the VO_{2max} of individuals use the relationship between heart rate and oxygen uptake.

The maximum oxygen uptake occurs when an individual's heart is beating at its maximum rate.

Graph 2 shows measurements of heart rate and oxygen uptake for a professional sportsman and an office worker, who are both 24 years old. The measurements were taken as speed was gradually increased on a treadmill.

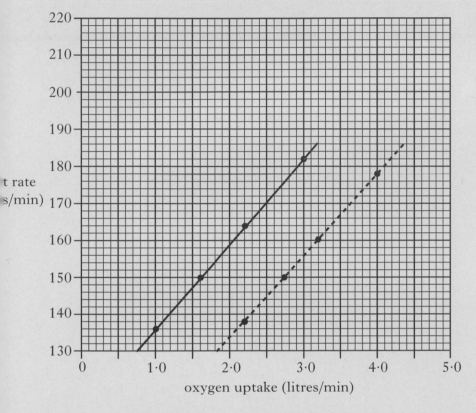

Key

——————— Office worker

- - - - - - - - Sportsman

heart rate
(beats/min)

oxygen uptake (litres/min)

(*d*)   (i)   An individual's maximum heart rate can be calculated by subtracting their age from 220.

Calculate the maximum heart rate of the office worker.

Space for calculation

_____ beats/min   **1**

(ii)   **Use the graph** to predict the maximum oxygen uptake of the office worker.

_____ litres/min   **1**

(iii)   The sportsman weighed 60 kg.

Use the information in **graphs 1** and **2** to determine his sport.

_____   **1**

Marks

8. The diagram below shows the human heart and some associated blood vessels. The arrows on the diagram show the direction of blood flow.

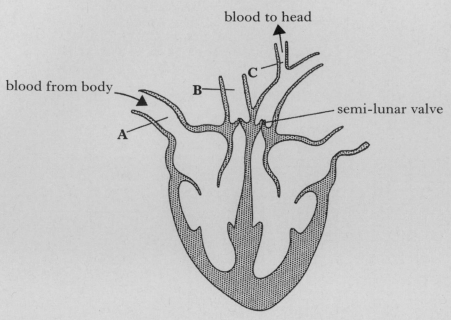

blood to head

blood from body

C

B

semi-lunar valve

A

(a) Name blood vessels **A**, **B** and **C**.

A _____

B _____

C _____   2

(b) Place arrows on the diagram to show the path of oxygenated blood as it flows through the heart.   1

(c) Describe the function of the semi-lunar valve **labelled on the diagram**.

_____

_____   1

(d) During which stage of the cardiac cycle do the semi-lunar valves open?

_____   1

Marks

9. (a) The diagram below shows a synapse which links a nerve cell with the sinoatrial node (SAN) in the heart.

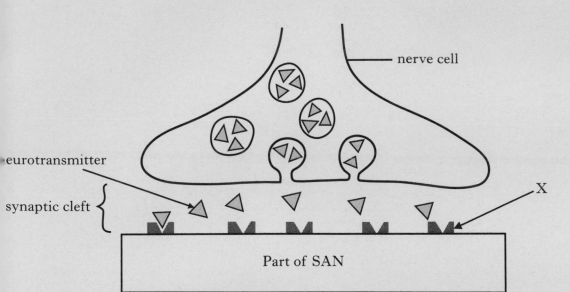

(i) Where in the heart is the SAN located?

_____   1

(ii) Describe the function of molecule X.

_____   1

(b) One example of a neurotransmitter is acetylcholine.

How is acetylcholine removed from the synapse?

_____

_____   1

(c) (i) In which area of the brain does the sympathetic nervous system originate?

_____   1

(ii) Describe a situation which would lead to stimulation of the sympathetic nervous system.

_____

_____   1

10. The diagram below shows two different neural pathways.

Nerve impulses are travelling from left to right in both pathways.

Marks

Pathway A

Pathway B

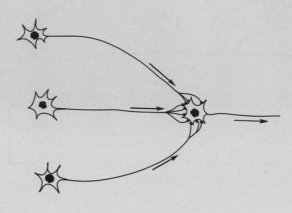

(*a*) (i) Name the types of pathway represented by **A** and **B**.

A _____

B _____   1

(ii) Pathway **A** helps the hand to function.

Explain how it does this.

_____

_____

_____   2

(*b*) Blinking is a reflex action.

(i) What is a reflex action?

_____

_____   1

(ii) The blinking reflex can sometimes be suppressed.

What term refers to the ability of the nervous system to suppress reflexes?

_____   1

11. One-fifth of all UK deaths are caused by smoking.

The graph below shows the total number of deaths from lung cancer of males and females of different ages in the United Kingdom in 2004.

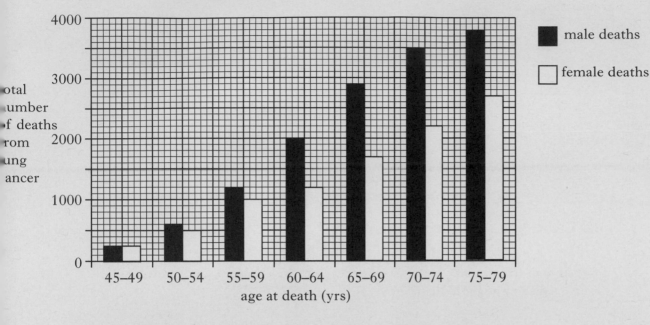

(*a*) Describe the **two** main trends shown by the graph.

1 _____

_____

2 _____

_____

2

(*b*) (i) Calculate the whole number ratio of male to female deaths in 45 to 49 year olds and 60 to 64 year olds.

Space for calculation

45–49 years _____ : _____   60–64 years _____ : _____
                        male   female   male   female

1

(ii) Suggest a reason for the **difference** between the two calculated ratios.

_____

_____

1

(*c*) Ninety-five percent of deaths from lung cancer occur in smokers.

Calculate how many male non-smokers aged 75 to 79 died from lung cancer in the UK in 2004.

Space for calculation

_____

1

Marks

12. The map below represents a short length of a Scottish river.

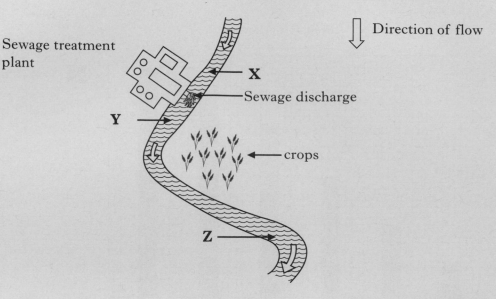

(a) The sewage treatment works sometimes gets overloaded and untreated sewage is discharged into the river.

(i) Following the discharge of sewage, state how bacteria would change in number between the following points.

Give a reason for your answer.

A Between points **X** and **Y**.

Change _____

Reason _____

_____   1

B Between points **Y** and **Z**.

Change _____

Reason _____

_____   1

(ii) State how algae would change in number between points **Y** and **Z**.

Give a reason for your answer.

Change _____

Reason _____

_____   1

Marks

12. **(continued)**

(b) Herbicides are frequently applied to land where crops are growing.

(i) What is a herbicide?

_____ 1

(ii) Explain how the use of herbicides leads to an increased crop yield.

_____

_____ 1

(c) Crop yield can be increased by the insertion of a gene from another organism into a chromosome of the crop plant.

Name this process.

_____ 1

[Turn over for Section C

Marks

SECTION C

Both questions in this section should be attempted.

Note that each question contains a choice.

Questions 1 and 2 should be attempted on the blank pages which follow.

Supplementary sheets, if required, may be obtained from the invigilator.

Labelled diagrams may be used where appropriate.

1. Answer **either** A **or** B.

    A. Discuss how other people can affect an individual's behaviour under the following headings:

      (i)   the influence of groups;   6

      (ii) influences that change beliefs.   4

                                                                          (10)

    OR

    B. Discuss global warming under the following headings:

      (i)   possible causes of global warming;   6

      (ii) potential effects of rising sea levels.   4

                                                                          (10)

In question 2, ONE mark is available for coherence and ONE mark is available for relevance.

2. Answer **either** A **or** B.

    A. Describe how immunity is naturally acquired.   **(10)**

    OR

    B. Describe the nature and reproduction of viruses.   **(10)**

[END OF QUESTION PAPER]

DO NOT
WRITE IN
THIS
MARGIN

SPACE FOR ANSWERS

SPACE FOR ANSWERS

DO NO WRITE THIS MARGI

SPACE FOR ANSWERS

DO NO
WRITE
THIS
MARGI

ADDITIONAL GRAPH FOR QUESTION 4(*e*)

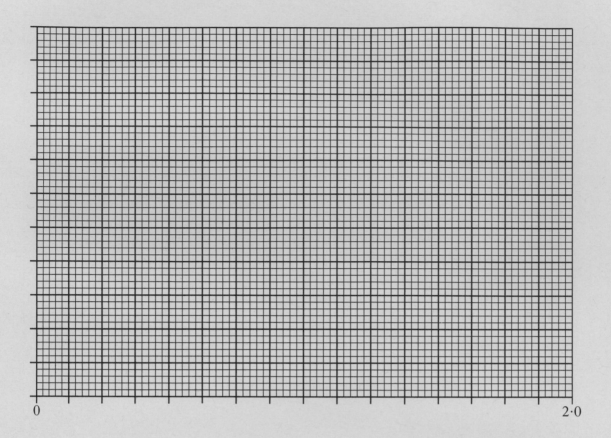

0   2·0

HIGHER

2010

[BLANK PAGE]

FOR OFFICIAL USE

Total for
Sections B & C

X009/301

NATIONAL
QUALIFICATIONS
2010

THURSDAY, 27 MAY
1.00 PM – 3.30 PM

HUMAN BIOLOGY
HIGHER

Fill in these boxes and read what is printed below.

Full name of centre

Town

Forename(s)

Surname

Date of birth

Day	Month	Year	Scottish candidate number	Number of seat

SECTION A—(30 marks)

Instructions for completion of Section A are given on page two.

For this section of the examination you must use an **HB pencil**.

SECTIONS B AND C—(100 marks)

1 (a) All questions should be attempted.

(b) It should be noted that in **Section C** questions 1 and 2 each contain a choice.

2 The questions may be answered in any order but all answers are to be written in the spaces provided in this answer book, **and must be written clearly and legibly in ink**.

3 Additional space for answers will be found at the end of the book. If further space is required, supplementary sheets may be obtained from the Invigilator and should be inserted inside the **front** cover of this book.

4 The numbers of questions must be clearly inserted with any answers written in the additional space.

5 Rough work, if any should be necessary, should be written in this book and then scored through when the fair copy has been written. If further space is required a supplementary sheet for rough work may be obtained from the Invigilator.

6 Before leaving the examination room you must give this book to the Invigilator. If you do not, you may lose all the marks for this paper.

Read carefully

1 Check that the answer sheet provided is for **Human Biology Higher (Section A)**.

2 For this section of the examination you must use an **HB pencil**, and where necessary, an eraser.

3 Check that the answer sheet you have been given has **your name**, **date of birth**, **SCN** (Scottish Candidate Number) and **Centre Name** printed on it.

Do not change any of these details.

4 If any of this information is wrong, tell the Invigilator immediately.

5 If this information is correct, **print** your name and seat number in the boxes provided.

6 The answer to each question is **either** A, B, C or D. Decide what your answer is, then, using your pencil, put a horizontal line in the space provided (see sample question below).

7 There is **only one correct** answer to each question.

8 Any rough working should be done on the question paper or the rough working sheet, **not** on your answer sheet.

9 At the end of the examination, put the **answer sheet for Section A inside the front cover of this answer book**.

Sample Question

The digestive enzyme pepsin is most active in the

A stomach

B mouth

C duodenum

D pancreas.

The correct answer is **A**—stomach. The answer **A** has been clearly marked in **pencil** with a horizontal line (see below).

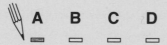

Changing an answer

If you decide to change your answer, carefully erase your first answer and, using your pencil, fill in the answer you want. The answer below has been changed to **D**.

SECTION A

All questions in this section should be attempted.

Answers should be given on the separate answer sheet provided.

1. The diagram below shows an enzyme-catalysed reaction taking place in the presence of an inhibitor.

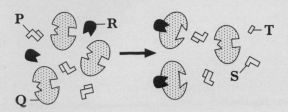

Which line in the table below identifies correctly the molecules in the reaction?

	Inhibitor	Substrate	Product
A	P	R	S
B	Q	P	S
C	R	P	T
D	R	Q	T

2. The following diagram shows a branched metabolic pathway.

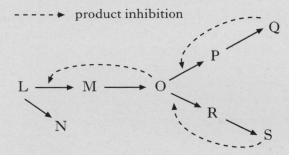

Which reaction would tend to occur if both Q and S are present in the cell in high concentrations?

A L → M

B R → S

C O → P

D L → N

3. A fragment of DNA was found to have 120 guanine bases and 60 adenine bases. What is the total number of sugar molecules in this fragment?

A   60

B   90

C   180

D   360

4. The following information refers to protein synthesis.

tRNA anticodon	amino acid carried by tRNA
G U G	Histidine (his)
C G U	Alanine (ala)
G C A	Arginine (arg)
A U G	Tyrosine (tyr)
U A C	Methionine (met)
U G U	Threonine (thr)

What order of amino acids would be synthesised from the base sequence of DNA shown?

Base sequence of DNA

C G T T A C G T G

A   arg - tyr - his

B   ala - met - his

C   ala - tyr - his

D   arg - tyr - thr

5. In which of the following is the cell organelle listed correctly with its function?

	Cell organelle	Function
A	Mitochondrion	Anaerobic respiration
B	Ribosome	Release of ATP
C	Lysosome	Synthesis of enzymes
D	Nucleolus	Synthesis of RNA

6. Carrier molecules involved in the process of active transport are made of

A   protein

B   carbohydrate

C   lipid

D   phospholipid.

[Turn over

7. An investigation was carried out into the uptake of sodium ions by animal cells. The graph shows the rates of sodium ion uptake and breakdown of glucose at different concentrations of oxygen.

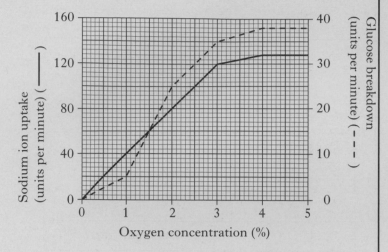

Calculate the number of units of sodium ions that are taken up over a 5 minute period when the concentration of oxygen in solution is 2%.

A   80

B   100

C   400

D   500

8. Which of the following statements about viruses is true?

A   Viral protein directs the synthesis of new viruses.

B   New viruses are assembled outside the host cell.

C   Viral protein is injected into the host cell.

D   Viral DNA directs the synthesis of new viruses.

9. What is the significance of chiasma formation?

A   It results in the halving of the chromosome number.

B   It results in the pairing of homologous chromosomes.

C   It permits gene exchange between homologous chromosomes.

D   It results in the independent assortment of chromosomes.

10. The transmission of a gene for deafness is shown in the family tree below.

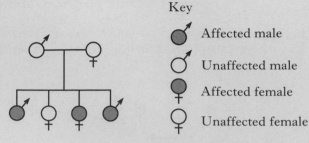

This condition is controlled by an allele which is

A   dominant and sex-linked

B   recessive and sex-linked

C   dominant and not sex-linked

D   recessive and not sex-linked.

11. The examination of a karyotype would **not** detect

A   phenylketonuria

B   Down's syndrome

C   the sex of the fetus

D   evidence of non-disjunction.

12. A woman with blood group AB has a child to a man with blood group O. What are the possible phenotypes of the child?

A A or B

B AB only

C AB or O

D AB, A or B

13. Cystic fibrosis is an inherited condition caused by a recessive allele. The diagram below is a family tree showing affected individuals.

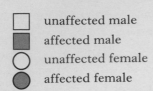

    ☐ unaffected male
    ■ affected male
    ○ unaffected female
    ● affected female

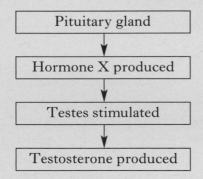

Which two individuals in this family tree must be heterozygous for the cystic fibrosis gene?

A   3 and 5

B   4 and 6

C   1 and 2

D   2 and 6

14. The diagram below shows the influence of the pituitary gland on testosterone production.

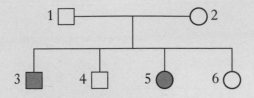

| Pituitary gland |
| Hormone X produced |
| Testes stimulated |
| Testosterone produced |

What is hormone X?

A   Luteinising hormone

B   Follicle stimulating hormone

C   Oestrogen

D   Progesterone

15. From which structure in the female reproductive system does a corpus luteum develop?

A   Endometrium

B   Graafian follicle

C   Fertilised ovum

D   Unfertilised ovum

16. The table below contains information about four semen samples.

	Semen sample			
	A	B	C	D
Number of sperm in sample (millions/cm^3)	40	30	20	60
Active sperm (percent)	50	60	75	40
Abnormal sperm (percent)	30	65	10	70

Which semen sample has the highest number of active sperm per cm^3?

17. Which of the following describes correctly the exchange of materials between maternal and fetal circulations?

	Glucose	Antibodies
A	into fetus by active transport	into fetus by active transport
B	into fetus by active transport	into fetus by pinocytosis
C	into fetus by pinocytosis	into fetus by active transport
D	into fetus by diffusion	into mother by pinocytosis

18. The diffusion pathway of carbon dioxide within body tissues is

A   plasma → tissue fluid → cell cytoplasm

B   lymph → tissue fluid → cell cytoplasm

C   cell cytoplasm → tissue fluid → plasma

D   tissue fluid → lymph → plasma.

[Turn over

19. The graph below shows changes in arterial blood pressure.

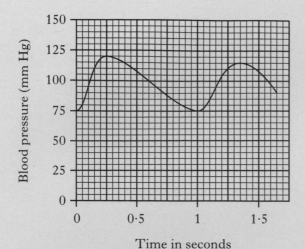

Time in seconds

The shape of the graph is due to

A   the action of the heart muscle

B   the action of the diaphragm

C   the closing of the valves in the veins

D   muscular contraction of the arteries.

20. An ECG trace is shown below.

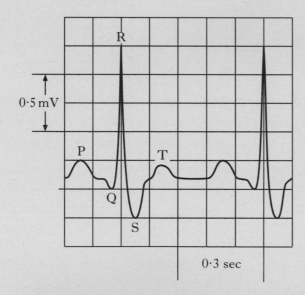

What is the person's heart rate?

A   100 beats per minute

B   120 beats per minute

C   150 beats per minute

D   200 beats per minute

21. Which of the following statements refers correctly to the cardiac cycle?

A   During systole the atria contract followed by the ventricles.

B   During systole the ventricles contract followed by the atria.

C   During diastole the atria contract followed by the ventricles.

D   During diastole the ventricles contract followed by the atria.

22. Which line in the table below correctly describes the conditions under which the affinity of haemoglobin for oxygen is highest?

	Oxygen tension	Temperature (°C)
A	high	40
B	high	37
C	low	37
D	low	40

23. Which of the following is triggered by the hypothalamus in response to an increase in the temperature of the body?

A   Contraction of the hair erector muscles and vasodilation of the skin arterioles

B   Contraction of the hair erector muscles and vasoconstriction of the skin arterioles

C   Relaxation of the hair erector muscles and vasodilation of the skin arterioles

D   Relaxation of the hair erector muscles and vasoconstriction of the skin arterioles

24. The graph below shows the rate of sweating of an individual in different environmental conditions.

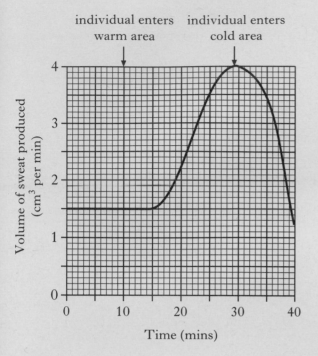

How long after entering the warm area did it take for the volume of sweat production to increase by 100%?

A   8 minutes

B   13 minutes

C   20 minutes

D   23 minutes

25. The diagram below shows the main parts of the brain as seen in vertical section.

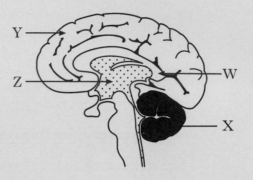

Which line in the table below correctly identifies the functions of two areas of the brain?

	Communication between hemispheres	Reasoning
A	W	X
B	X	Y
C	W	Y
D	Z	W

[Turn over

26. The diagram below shows a test on a man who had a damaged corpus callosum. This meant that he could no longer transfer information between his right and left cerebral hemispheres.

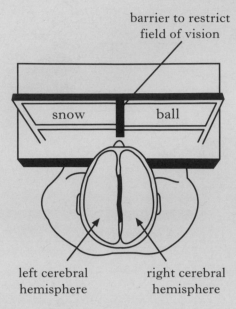

Some of the functions of each hemisphere are described in the table below.

Left cerebral hemisphere	Right cerebral hemisphere
processes information from right eye	processes information from left eye
controls language production	controls spatial task co-ordination

The man was asked to look straight ahead and then the words "snow" and "ball" were flashed briefly on the screen as shown.

What would the man say that he had just seen?

A   Ball

B   Snow

C   Snowball

D   Nothing

27. Which of the following statements about diverging neural pathways is correct?

A   They accelerate the transmission of sensory impulses.

B   They suppress the transmission of sensory impulses.

C   They decrease the degree of fine motor control.

D   They increase the degree of fine motor control.

28. Which of the following describes the change in an individual's behaviour where the presence of others causes the individual to show less restraint and become more impulsive?

A   Social facilitation

B   Shaping

C   Generalisation

D   Deindividuation

29. Which of the following identifies correctly a process in the nitrogen cycle?

A   Nitrifying bacteria trap atmospheric nitrogen.

B   Nitrifying bacteria convert ammonium compounds to nitrates.

C   Nitrogen-fixing bacteria convert nitrates to atmospheric nitrogen.

D   Denitrifying bacteria convert ammonia to nitrates.

30. The diagrams below contain information about the population of Britain.

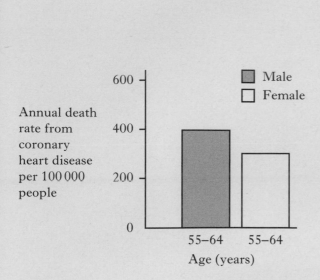

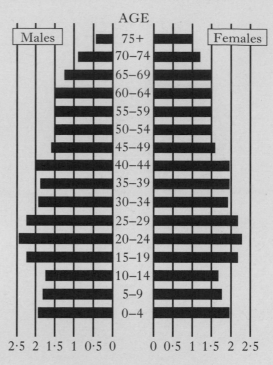

Population size (millions)

How many British men between 55 and 64 years of age die from coronary heart disease annually?

A   400

B   6000

C   12 000

D   24 000

**Candidates are reminded that the answer sheet MUST be returned
INSIDE the front cover of this answer booklet.**

[Turn over for Section B on *Page eleven*

[BLANK PAGE]

Marks

SECTION B

All questions in this section should be attempted.

All answers must be written clearly and legibly in ink.

1. The diagram below represents stages in the production of human sperm.

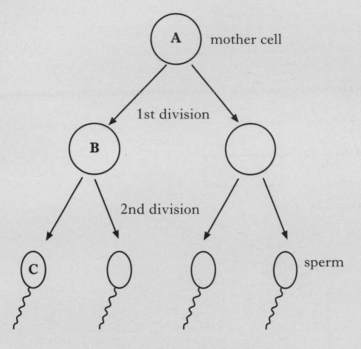

(a) Name the type of cell division that produces sex cells.

_____

1

(b) State the number of chromosomes which would be present in the cells labelled A, B and C.

A: _____ B: _____ C: _____

1

(c) Compare the appearance of the chromosomes in cell B and cell C.

_____

_____

1

(d) Name the **two** processes which increase variation during the 1st division of the sperm mother cell.

1 _____

2 _____

1

(e) State the location of sperm production in the testes.

_____

1

DO NO
WRITE
THI
MARG

Marks

2.   The diagram below shows some of the reactions which occur during aerobic respiration.

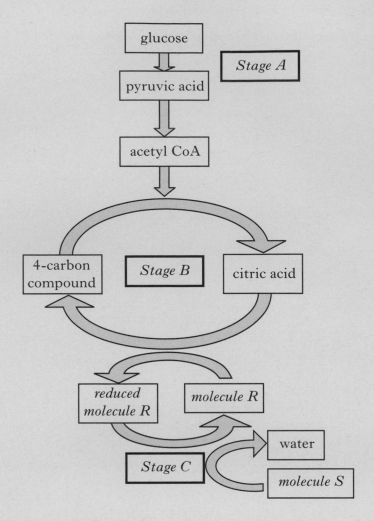

(a)   Complete the table by naming stages A, B and C and indicating their **exact** location within the cell.

Stage	Name	Location
A		
B		
C		

3

(b)   A glucose molecule contains 6 carbon atoms.

How many carbon atoms are found in the following molecules?

Pyruvic acid _____

Citric acid _____

1

Marks

2.   **(continued)**

(c)   Complete the following sentences by naming molecules R and S and describing their function with respect to stage C.

R is _____ and its function is _____

_____

_____ .

S is _____ and its function is _____

_____

_____ .

2

(d)   Under normal circumstances carbohydrate is the main respiratory substrate.

In each of the following extreme situations, state the alternative respiratory substrate and explain why the body has to use it.

Situation	Respiratory substrate	Explanation
Prolonged starvation		
Towards the end of a marathon race		

2

[Turn over

Marks

3. The diagram below shows blood from a person who has been infected by bacteria. These bacteria have triggered an immune response involving proteins P and Q.

The diagram is not drawn to scale.

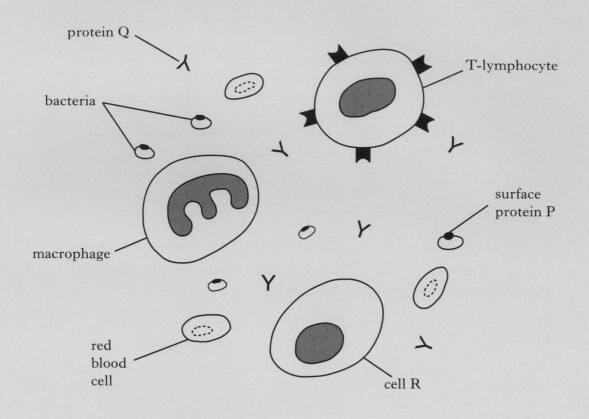

(a) (i) Identify proteins P and Q.

P _____   Q _____   1

(ii) Cell R produced protein Q.

Name this type of cell.

_____   1

(iii) Describe the role of the following cells in combating infection.

(A) T-lymphocyte _____

_____   1

(B) Macrophage _____

_____   1

Marks

3. **(continued)**

(b) Complete the following sentences by <u>underlining</u> one option from each pair of options shown in **bold**.

(i) Immunity gained after contracting a bacterial infection is an example of **active** / **passive** immunity that is **naturally** / **artificially** acquired.

1

(ii) Immunity gained from the injection of a tetanus vaccine is an example of **active** / **passive** immunity that is **naturally** / **artificially** acquired.

1

(c) What happens during an autoimmune response?

_____

_____ **1**

[Turn over

DO NO
WRITE
THIS
MARG

Marks

4. Lactose is the main sugar found in milk.

Lactose is broken down by lactase, an enzyme which is made by cells lining the small intestine. The glucose and galactose molecules produced are then absorbed into the bloodstream.

$$\text{lactose} \xrightarrow{\text{lactase}} \text{glucose} + \text{galactose}$$

A student carried out an investigation to compare the lactose content of human milk and cow milk.

He set up a test tube containing human milk and lactase solution. Every 30 seconds samples were taken and the glucose concentration measured. Then he repeated the procedure with cow milk.

His experimental setup is shown in Figure 1.

His results are shown in the table below.

Time (min)	Concentration of glucose (%)	
	Human milk	Cow milk
0	0	0
0·5	0·28	0·28
1·0	0·54	0·46
1·5	0·80	0·54
2·0	1·04	0·58
2·5	1·10	0·58
3·0	1·10	0·58

Figure 1

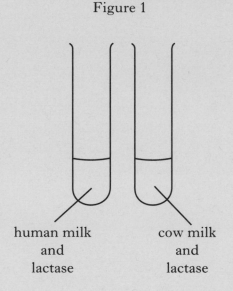

human milk
and
lactase

cow milk
and
lactase

(*a*) Lactose is a disaccharide sugar.

Explain how the information above supports this statement.

_____

_____   1

(*b*) One variable that must be kept constant in this investigation is pH.

List **two** other variables which would have to be kept constant.

1 _____

2 _____   1

Marks

4. **(continued)**

(c) Construct a line graph to show all the data in the table.

(Additional graph paper, if required, can be found on *Page thirty-six.*)

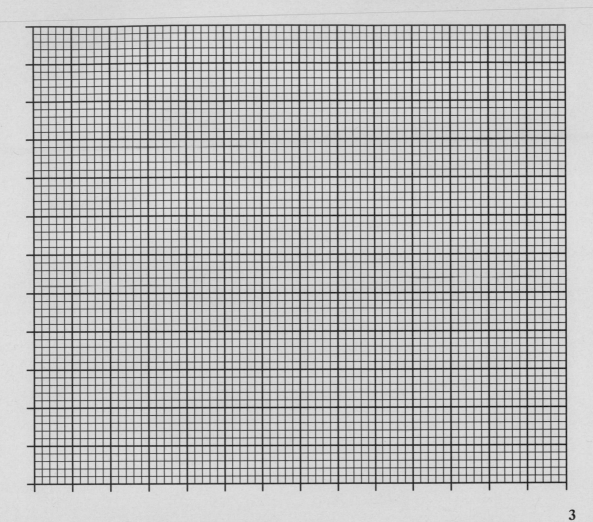

3

(d) What conclusion can be drawn from this investigation?

_____

_____ 1

(e) Suggest a reason why the rate of glucose production is not constant throughout the investigation.

_____

_____ 1

(f) How could the student improve the reliability of his results?

_____

_____ 1

DO NO
WRITE
THIS
MARG

Marks

4. **(continued)**

(g) Some people who have problems digesting lactose are said to be lactose intolerant.

They cannot produce the enzyme lactase.

(i) What general phrase describes an inherited disorder in which the absence of an enzyme prevents a chemical reaction from happening?

_____   1

(ii) A test can be carried out for lactose intolerance.

The individual being tested does not eat for twelve hours and then drinks a liquid that contains lactose. The individual rests for the next two hours during which their blood glucose level is measured at regular intervals.

What results would be expected if the individual is lactose intolerant?

_____

_____   1

Marks

5. The diagram below shows a section of a woman's breast shortly after she has given birth.

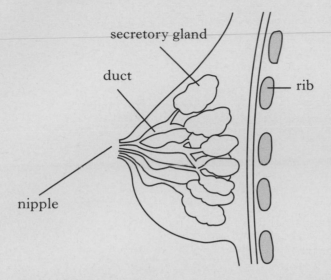

secretory gland

duct

rib

nipple

(a) (i) Name the hormone that stimulates the secretory glands to start producing milk.

_____

1

(ii) The cells lining the secretory glands are particularly rich in ribosomes. Suggest a reason for this.

_____

_____

1

(b) Fluid is not usually released from the breast until the baby suckles.

(i) What name is given to the first fluid that the baby receives from the breast?

_____

1

(ii) Describe **one** way in which this first fluid differs from the breast milk produced a few days later.

_____

_____

1

(iii) Suckling and crying are examples of non-verbal communication used by a baby. Why is non-verbal communication important to **both** the mother and baby?

_____

_____

1

DO NC
WRITE
THIS
MARG

Marks

6. The flow diagram below summarises what happens in the body after a meal of fish and chips.

```
┌─────────────────────────────────┐
│  Digestion of fish and chips in the │
│    stomach and small intestine   │
└─────────────────────────────────┘
                 │
                 ▼
┌─────────────────────────────────────┐
│ Absorption of the products of digestion through the │
│         walls of the small intestine          │
└─────────────────────────────────────┘
                 │
                 ▼
┌─────────────────────────────┐
│   Metabolism of some absorbed   │
│     substances by the liver     │
└─────────────────────────────┘
                 │
                 ▼
┌─────────────────────────────────────┐
│ Transport of some products of metabolism │
│     around the body in the bloodstream     │
└─────────────────────────────────────┘
```

(a) Explain how bile salts aid the digestion of the fish and chips.

_____

_____ 1

(b) The products of fat digestion are fatty acids and glycerol.

Describe the route taken by these products as they move from the small intestine to the bloodstream.

_____

_____

_____

_____ 2

Marks

6. (continued)

(c) During the absorption and metabolism of this meal, samples of blood from the hepatic portal vein and the hepatic vein were tested for glucose and urea.

Complete each row of the table below, using the words **Higher** and **Lower**, to compare the concentration of each substance in the two blood vessels.

	Blood vessel	
Substance	Hepatic portal vein	Hepatic vein
Glucose		
Urea		

2

(d) State **one** feature of veins which helps to maintain blood flow.

_____

_____ 1

(e) Drugs and alcohol pass into the bloodstream through the digestive system.

The liver converts these harmful substances into harmless products.

What term describes this action of the liver?

_____ 1

[Turn over

Marks

7. A long distance runner took part in some laboratory tests using a treadmill.

She was asked to use the treadmill at a setting of 4 km/h for three minutes during which her pulse rate was monitored. At the end of this time a blood sample was taken which was tested for lactic acid concentration. The procedure was then repeated a number of times at faster speeds.

The results of the tests are shown in the graph below.

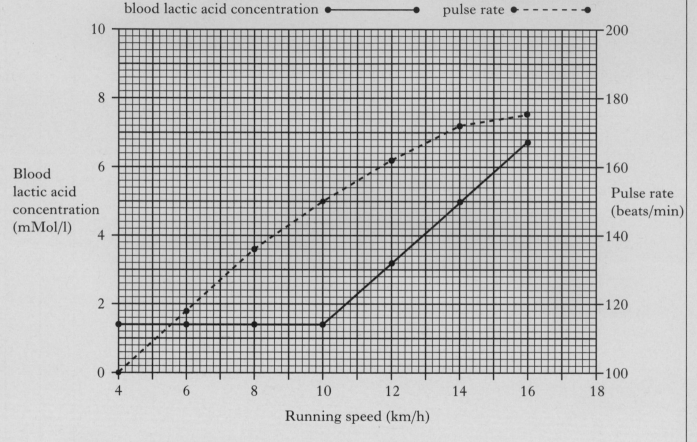

(*a*) (i) What was the runner's pulse rate when she was running at 6 km/h?

_____ **1**

(ii) State the concentration of lactic acid in the runner's blood when her pulse rate was 172 beats/min.

_____ mMol/l **1**

(iii) Predict what the runner's blood lactic acid concentration would be if she ran at 18 km/h for three minutes.

Blood lactic acid concentration _____ mMol/l **1**

DO NOT
WRITE IN
THIS
MARGIN

Marks

7. **(continued)**

(b) A build-up of lactic acid in muscles leads to fatigue.

(i) Explain why lactic acid builds up in the muscles as running speeds increase.

_____

_____

_____

_____ 2

(ii) Distance runners often monitor their pulse rate while they are training.

Suggest how this runner could use a pulse rate monitor and the information from the graph to allow her to run for long periods of time without developing muscle fatigue.

_____

_____

_____

_____ 2

[Turn over

DO NO
WRITE
THIS
MARG

Marks

8. Two men (P and R) were being tested for *diabetes mellitus*, a condition which results in failure to control blood glucose concentration.

After fasting overnight, they were given a large glucose drink. Their blood glucose concentration was measured immediately (0 hours) and then every hour for five hours.

The results of the tests are shown in the table below.

	Time after drinking glucose (hours)					
	0	*1*	*2*	*3*	*4*	*5*
Blood glucose concentration of P (mg/100 ml)	145	210	190	180	170	160
Blood glucose concentration of R (mg/100 ml)	90	125	90	85	90	90

(*a*)   It was concluded that P had diabetes and R did not.

(i)   State **two** ways in which the test results indicate that P has diabetes.

1 _____

2 _____   1

(ii)   Name the hormone responsible for the change in the blood glucose concentration of R

(A)   between 1 and 2 hours   _____

(B)   between 3 and 4 hours.   _____   1

(*b*)   *Diabetes insipidus* can be caused by a lack of ADH in the body.

(i)   Which organ of the body releases ADH?

_____   1

(ii)   State an effect that failure to produce ADH would have on the body.

_____

_____   1

Marks

9. The diagram below shows a synapse between two nerve cells in the brain and a magnified view of a receptor called NMDA.

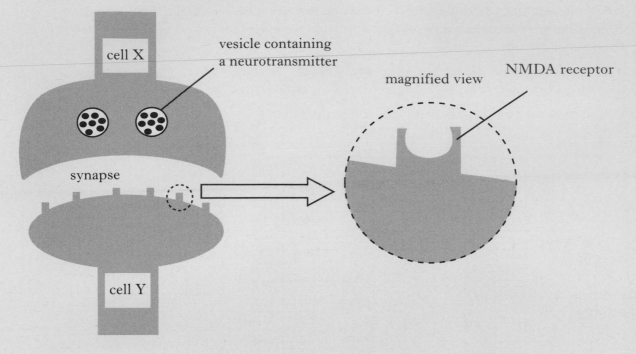

(a) (i) Describe how the neurotransmitter in the vesicle reaches cell Y.

_____

_____

_____   2

(ii) The diagram above shows a single neural pathway.

Explain how a converging neural pathway would be more likely to generate an impulse in nerve cell Y.

_____

_____

_____   2

(b) Many factors can lead to memory loss.

(i) One of these factors is a reduction in the number of NMDA receptors.

Which part of the brain contains nerve cells rich in NMDA receptors?

_____   1

(ii) Another factor is the decreased production of acetylcholine.

Name the condition which results from the loss of acetylcholine-producing cells in the brain.

_____   1

Marks

10. A study was carried out to compare the influence of genetics with that of the environment on the development of two behavioural conditions, A and B.

Several hundred pairs of children, from the same families, took part in the study. Some pairs were monozygotic twins, some pairs were dizygotic twins and some pairs were adopted and unrelated.

In each pair, one of the children had one of the behavioural conditions and investigators observed whether or not the other child shared the condition.

Results of the study are shown in the bar graph below.

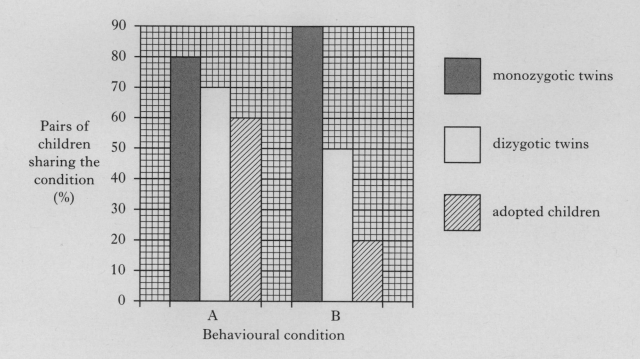

(a) Explain why it was important that monozygotic twins were chosen for this study.

_____

_____

_____   2

(b) Use the graph to explain whether conditions A and B are more likely to be caused by genetic or environmental factors.

   (i) Likely cause of condition A _____

        Explanation _____

        _____   1

   (ii) Likely cause of condition B _____

        Explanation _____

        _____   1

Marks

11. The bar graph shows population changes in Scotland for different age groups between 1991 and 2000.

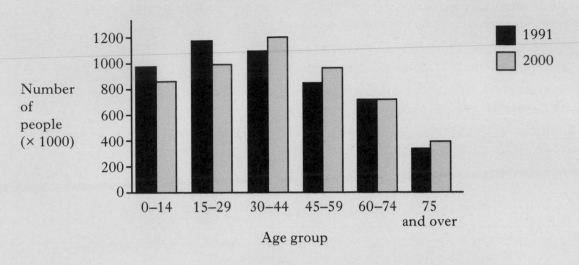

(*a*) Suggest a reason for the population change in those aged 75 and over.

_____

_____   1

(*b*) Describe **two** ways in which the data for the year 2000 would be different if it were taken from a developing country with a similar population size to Scotland.

1 _____

_____

2 _____

_____   1

(*c*) Describe **two** ways in which the information in the graph could be used by authorities to plan for the future.

1 _____

_____

2 _____

_____   1

[Turn over

Marks

12. An investigation was carried out into the influence of adults on the behaviour of young children.

Some groups of children watched a recording of either a man or a woman being physically and verbally aggressive to a large plastic clown.

Other groups of children watched either a man or a woman behaving in a non-aggressive manner towards the clown.

Each child was then placed in a room on their own with the clown. The number of aggressive acts they committed over a five minute period was counted.

The figures in the table below show the average number of aggressive acts that the children committed while in the room.

Gender of children	Average number of aggressive acts committed by the children			
	Aggressive man observed	Aggressive woman observed	Non-aggressive man observed	Non-aggressive woman observed
Boys	18·7	7·9	1·0	0·6
Girls	4·4	9·2	0·2	0·8

(a) (i) Which adult/child combination resulted in the least aggression?

_____   1

(ii) Calculate the percentage increase in aggressive acts committed by boys when they observe an aggressive man rather than a non-aggressive man.

Space for calculation

_____ %   1

(iii) State a conclusion that can be drawn from these results regarding the gender of the aggressive adult.

_____

_____   1

(b) The children are observing and then repeating the acts of adults. What form of learning are they using?

_____   1

(c) Suggest a control that could have also been used in this investigation.

_____

_____   1

13. The graph below shows the application rates of nitrogen and phosphorus to crops in an area of Scotland between 1986 and 2006.

Marks

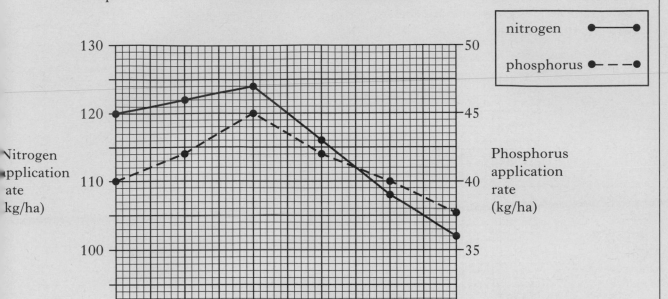

(a) Describe **one** similarity and **one** difference in the data for nitrogen and phosphorus application rate between 1986 and 2006.

Similarity _____

_____

Difference _____

_____ **2**

(b) Express, as a simple whole number ratio, the application rate of nitrogen compared to phosphorus in 1986.

Space for calculation

_____ : _____ **1**
nitrogen   phosphorus

(c) In recent years, there has been a decrease in the use of nitrogen and phosphorus on farms in Scotland.

(i) Suggest **one** way in which this decrease might benefit the environment.

_____

_____ **1**

(ii) Suggest **one** way in which this decrease might disadvantage farmers.

_____

_____ **1**

DO N
WRIT
THI
MAR(

Marks

14. Glaciers are large masses of ice on mountains and in cold regions of the world. The graph below shows the average change in glacier thickness around the world between 1955 and 2005.

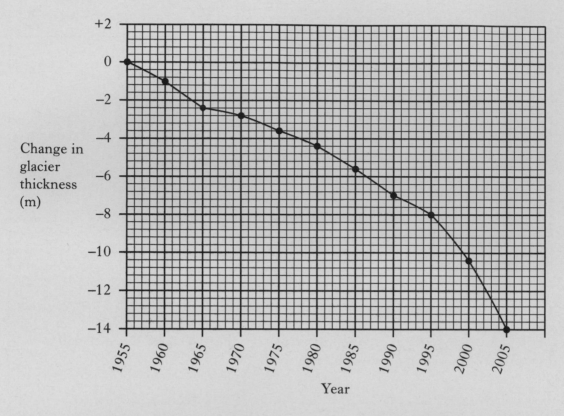

(*a*) (i) Calculate the average yearly decrease in glacier thickness between 1955 and 2005.

Space for calculation

_____ m/year   **1**

(ii) One consequence of this decrease in glacier thickness is rising sea levels. Describe **one** effect of rising sea levels and subsequent flooding on coastal communities around the world.

_____

_____   **1**

Marks

14. (continued)

(b) Many people believe that the change in glacier thickness is caused by global warming.

   (i) Name **two** gases that contribute to global warming.

   1 _____ 2 _____   **1**

   (ii) Give **two** reasons why one of these gases is increasing in the atmosphere.

   Gas _____

   Reason 1 _____

   _____

   Reason 2 _____

   _____   **1**

[Turn over for Section C on *Page thirty-two*

Marks

SECTION C

Both questions in this section should be attempted.

Note that each question contains a choice.

Questions 1 and 2 should be attempted on the blank pages which follow.

Supplementary sheets, if required, may be obtained from the Invigilator.

Labelled diagrams may be used where appropriate.

1. Answer **either** A **or** B.

   A. Discuss memory under the following headings:

      (i) short-term memory;   **5**

      (ii) the transfer of information between short and long-term memory.   **5**

   OR   **(10)**

   B. Discuss how man has attempted to increase food supply under the following headings:

      (i) chemical use;   **4**

      (ii) genetic improvement;   **3**

      (iii) land use.   **3**

   (10)

In question 2, ONE mark is available for coherence and ONE mark is available for relevance.

2. Answer **either** A **or** B.

   A. Discuss the biological basis of contraception.   **(10)**

   OR

   B. Discuss the conducting system of the heart and how it is controlled.   **(10)**

[END OF QUESTION PAPER]

DO NOT
WRITE IN
THIS
MARGIN

SPACE FOR ANSWERS

SPACE FOR ANSWERS

SPACE FOR ANSWERS

DO NOT
WRITE IN
THIS
MARGIN

SPACE FOR ANSWERS

DO NOT
WRITE IN
THIS
MARGIN

SPACE FOR ANSWERS

ADDITIONAL GRAPH PAPER FOR QUESTION 4(*c*)

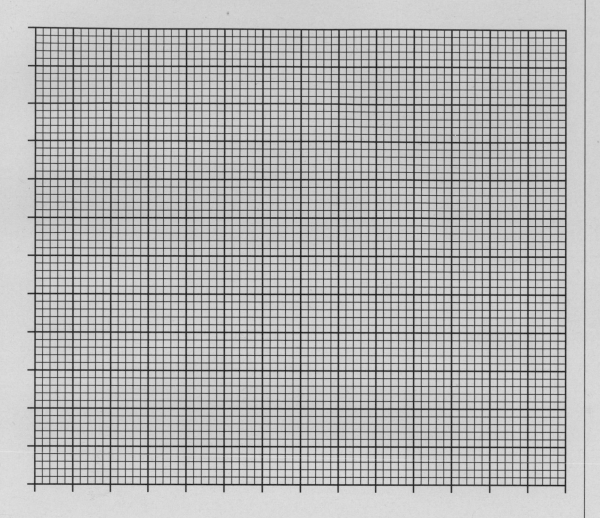